BIOFORMATION OF FLAVOURS

Bioformation of Flavours

Edited by
R.L.S. Patterson
Department of Animal Husbandry, University of Bristol

B.V. Charlwood
Department of Biology, King's College, London

G. MacLeod
Department of Food and Nutritional Science, King's College, London

A.A. Williams
Sensory Research Laboratories Ltd., Nailsea, Bristol

ROYAL
SOCIETY OF
CHEMISTRY
Information
Services

The proceedings of an International Conference organized jointly by the Food Chemistry Group of the Royal Society of Chemistry and the Phytochemical Society of Europe, King's College, London, 19–21 December 1990.

Special Publication No. 95

ISBN 0-85186-446-5

A catalogue record for this book is available from the British Library

Published by The Royal Society of Chemistry,
Thomas Graham House, Science Park, Cambridge
CB4 4WF

Printed and bound in Great Britain by Bookcraft (Bath) Ltd.

Preface

An International Conference, entitled 'Bioformation of Flavours' was organized jointly by the Food Chemistry Group of the Royal Society of Chemistry and the Phytochemical Society of Europe in December 1990. The aim of the conference was to gather together researchers working in various fields of the chemistry of naturally-occurring flavour compounds, to discuss progress and to provide a forum for the presentation of the latest results in a number of associated areas under the general terms of 'production and improvement of natural food flavours'.

The papers presented at the conference are brought together in this volume and cover current concepts of flavour production by plants, fungi, yeasts, and bacteria, as well as more specific aspects of flavour creation and flavour improvement that may be achieved by use of modern *in vitro* and *in vivo* techniques. Topics include: improvements in flavour attained through plant breeding, plant cell culture, and microbial selection; the biosynthesis of plant flavour compounds; biopathways to novel flavours; chiral flavour compounds, and bioreactors for the industrial production of flavours from plant cells, micro-organisms, and enzymes. The proceedings concluded with papers on the problems of scale-up in novel systems and on concentration and isolation techniques used in downstream processing.

The range of topics covered is wide and the book should have appeal for those working in natural product chemistry, synthetic organic chemistry, chromatography and mass spectrometry, plant breeding, plant physiology, cell culture techniques, and in applied microbiology; also to those concerned with the technical and commercial aspects of natural product production in industry. Above all, however, it is intended for both the industrial and university research worker who is keen to keep abreast of developments in this challenging and exciting field.

Contents

Bioflavours: An Overview

P. Schreier

LEHRSTUHL FÜR LEBENSMITTELCHEMIE, UNIVERSITÄT
WÜRZBURG, AM HUBLAND, 8700 WÜRZBURG, GERMANY

1 Introduction

Three years ago, the term 'bioflavour' was used for the first time as the title
of an international symposium at Würzburg University.[1] In his introductory
lecture, Drawert[2] defined 'bioflavours' as (i) 'natural' and (ii) 'naturally produced' flavours. In the following chapter, an attempt is made to give answers
to the following questions: (i) when is a flavouring natural? (ii) what are the
reasons for the interest in natural flavours? (iii) what are the possibilities to
produce natural flavours? and (iv) how can they be differentiated analyti-
cally from nature-identical ones?

1.1 Natural Flavours

Legislation for the utilization of flavourings was recently discussed thor-
oughly during the course of an SCI Flavour and Fragrance Symposium. As
outlined in the abstract of a relevant contribution, in an attempt to harmon-
ize legislation before 1992, the Commission of the European Community
has brought out a framework directive, which may serve to complicate the
issues further.[3] Regarding the definition of 'natural', it seems to be easier to
refer to the US Federal Register, in which the term 'natural flavour' or 'natural flavouring' means an essential oil; oleoresin; essence or extractive; protein hydrolysate; or distillate; or any product of roasting, heating, or
enzymolysis, which contains the flavouring constituents derived from a
spice, fruit or fruit juice, vegetable or vegetable juice, edible yeast, herb, bark,
bud, root, leaf or similar plant material, meat, seafood, poultry, eggs, dairy

[1] 'Bioflavour '87', ed. P. Schreier, W. de Gruyter, Berlin, New York, 1988.
[2] F. Drawert, in 'Bioflavour '87', ed. P. Schreier, W. de Gruyter, Berlin, New York, 1988, p.3.
[3] J. Hardinge, *Chem. Ind. (London)*, 1990, 694.

products, or fermentation products thereof, whose significant function in food is flavouring rather than nutritional.

Thus, the definition of 'natural' is broad and the fine points are subject to legal interpretation. Most important from the standpoint of natural flavour production is that products derived by enzymic action and fermentation are considered to be 'natural' substances. The same regulation has been set in the recent EC directive.[4]

2   Biotechnology and Flavour Production

Since early times, flavour compounds ranging from complex mixtures to single substances have been extracted from plant sources or produced for flavours by fermentation. As the structures of the compounds began to be elucidated, more and more synthetic flavours became available. In spite of these advances, many flavour compounds are still extracted from botanical sources for economic reasons and because of the consumer's preference for natural products.[5] In recent years, the market has undergone a tremendous 'back-to-nature' demand,[6] which is probably due to an increased incidence of non-natural chemicals being reported to exhibit some form of toxicity. As a consequence, consumers may feel more at ease with 'natural' compounds and reject 'artificial' ones.[7]

With the increasing interest in natural products, more pressure has been directed to the production of natural flavours from raw materials such as plants.[8] These raw materials, however, are subject to various problems. These disadvantages are not discussed here in detail. Rather, the advantages of alternative methodologies, *i.e.* biotechnological processes, are demonstrated, *i.e.* (i) the products may possess the legal status of a 'natural' substance; (ii) defined stereochemistry is guaranteed by the high substrate and reaction specificity of biocatalysts; (iii) multiple step reactions proceed under mild conditions; (iv) adverse external influences, such as unfavourable climate, pest infestations, economical or ecological draw-backs, can be neglected.

As outlined in Figure 1, biotechnology[1, 9-15] comprises biotransformation,

[4] Richtlinie des Rates vom 22.6.1988 zur Angleichung der Rechtsvorschriften der Mitgliedsstaaten über Aromen zur Verwendung in Lebensmitteln und über Ausgangsstoffe für ihre Herstellung. Amtsblatt der Europäischen Gemeinschaften Nr. L 184/61 (15.7.1988).
[5] J.D. Dziezak, *Food Technol.*, 1986, **40**, 108.
[6] M.H. Tyrrell, *Food Technol.*, 1990, **44**, 68.
[7] J. Stofberg, in 'Biogeneration of Aromas', ed. T.H. Parliment and R. Croteau, ACS Symposium Series No. 317; American Chemical Society, Washington DC, 1986, p. 2.
[8] R.J. Whitaker and D.A. Evans, *Food Technol.*, 1987, **41**, 86.
[9] 'Food Biotechnology', ed. D. Knorr, Dekker, New York, 1987.
[10] P. Schreier, in 'Proceedings of the 8th International Biotechnology Symposium', ed. G. Durand, L. Bobichon, and J. Florent, Soc. Franc. Microbiol., Paris, 1988, p. 869.
[11] 'The Impact of Chemistry on Biotechnology', ed. M. Phillipps, S.P. Shoemaker, R.D. Middlekauff, and R.M. Ottenbrite, ACS Symposium Series No. 362, American Chemical Society, Washington DC, 1988.
[12] 'Biocatalysis in Agricultural Biotechnology', ed. J.R. Whitaker and P.E. Sonnet, ACS Symposium Series No. 389, American Chemical Society, Washington DC, 1989.

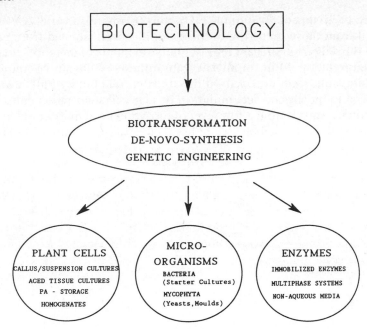

Figure 1   *Biotechnology and flavour production*

de-novo synthesis and genetic engineering. In the following, these three major areas are briefly discussed with regard to plant cells, microbial flavour production, flavour formation by enzymes, and research done in the field of genetic engineering. In addition, analytical aspects of the differentiation of 'natural' and 'nature-identical' flavours are considered.

2.1   Plant Tissue and Cell Culture

A large number of various fine chemicals is derived from plants, *e.g.* drugs, pigments, enzymes, and other biologically active substances. In the past, their production by means of plant cell cultures has attracted the interest of many researchers. As to flavour compounds from plant origin, comprehensive reviews have been provided recently.[16, 17] Although most cultures have been unable to produce an adequate yield of flavour substances, a few examples exist where tissue cultures have exhibited an increased production

13 P. Schreier, *Food Rev. Intern.*, 1989, **5**, 289.
14 'Biotechnology and the Food Industry', ed. P.L. Rogers and G.H. Fleet, Gordon and Breach Sci. Publ., London, New York, 1989.
15 'Biotechnology — Challenges for the Flavour and Food Industry', ed. R.C. Lindsay and B.J. Willis, Elsevier Applied Science, London, New York, 1989.
16 I. Koch-Heitzmann and W. Schultze, *Biochem. Physiol. Pflanzen*, 1989, **184**, 3.
17 T. Mulder-Krieger, R. Verpoorte, A. Baerheim-Svendsen, and J.J.C. Scheffer, *Plant Cell Tissue and Organ. Culture*, 1988, **13**, 85.
18 H. Sugisawa and Y. Ohnishi, *Agric. Biol. Chem.*, 1976, **40**, 231.
19 R. Takeda and K. Katoh, *Planta*, 1981, **151**, 525.
20 C. Violon, W. Sonck, and A. Vercruysse, *J. Chromatogr.*, 1984, **288**, 474.

compared with that of the plant.[18-20] On the other hand, a variety of volatiles different from those occurring in the plant has been isolated from cell cultures.[21, 22] In Figure 2, a recent representative example of metabolic reactions limiting product yield in the *de-novo* biosynthesis of flavour compounds in suspension cultures is outlined. In the study carried out by Falk *et al.*,[23] the absence of monoterpene accumulation in sage cell suspension cultures has been shown to result from a low level of biosynthetic activity coupled with a pronounced ability to catabolize camphor *via* the pathways outlined in Figure 2.

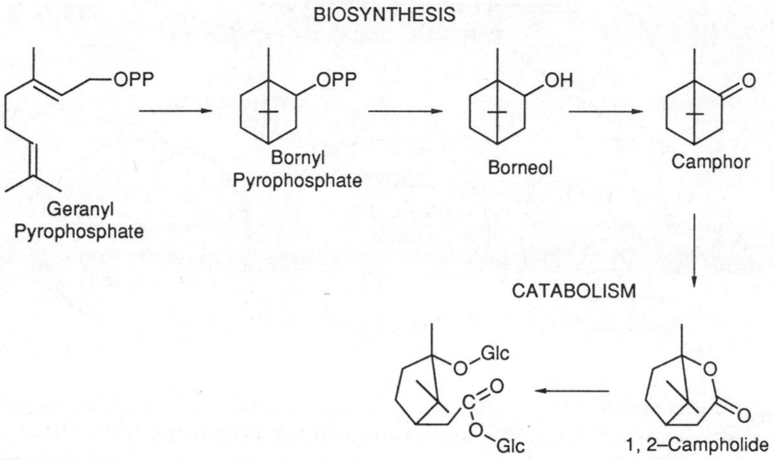

Figure 2 *Pathways for the conversion of geranyl pyrophosphate to camphor, and for the conversion of camphor to the glucoside–glucose ester of* 1,2-*campholide*[23]

As *de-novo* biosynthesis has been found unsuccessful in most cases, biotransformation of added precursors has been studied extensively. A comprehensive review has been provided recently.[24] Suspension cultures are often able to perform particular transformations. For instance, cofactor dependent specific conversions of terpenes in suspension cultures of aromatic plants often proceed with high yields and negligible amounts of by-products. Not only oxidations or reductions, as shown in Figure 3, but also numerous other types of reactions have been described.[24] Among them, glycosylation takes a particular place. Various plant cell cultures are capable of glycosylating exogenously administered aromatic compounds. In addition, the biotechnological productions of glycosides, to be used in the pharmaceutical or food industry, may be suggested.[2]

Summarizing this section, it has to be pointed out that the accumulation of larger amounts of flavour compounds in plant cell culture will continue to be a challenging scientific problem. As to media costs, in general,

[21] D.N. Butcher and J.D. Connolly, *J. Expert. Bot.*, 1971, **22**, 314.
[22] J. Berlin, *Chem. unserer Zeit.* 1983, **17**, 77.
[23] K.L. Falk, J. Gershenzon, and R. Croteau, *Plant Physiol.*, 1990, **93**, 1559.
[24] T. Suga and T. Hirata, *Phytochemistry*, 1990, **29**, 2393.

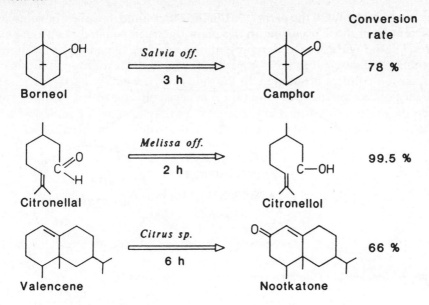

Figure 3 *Biotransformation of terpenes by* in-vitro *plant cell cultures*[2]

biotechnology using plant cell culture is regarded to be less favourable than other technologies, *e.g.* use of micro-organisms (Table 1).[25]

As an alternative, the concept of so-called precursor-stimulated flavour formation has to be discussed (Figure 4). It consists of the use of (i) aged tissue cultures, (ii) precursor atmosphere storage, and (iii) homogenates.[2] Of these three possibilities, only the latter will be discussed in the following.

Homogenates of plant tissues have to be considered as potential sources of natural flavours. In this area, special interest is devoted to the natural 'green' notes, as found in fresh homogenates of several leaves or apple peel. In spite of our knowledge about the fundamental steps of 'green odour' formation in homogenized plant tissues,[26] *i.e.* the biogeneration of hexanal and

Table 1 *Comparison of different parameters of microbial and plant cell biotechnology*[25]

	Microbial Cell culture	*Plant*
shear	insensitive	sensitive
doubling time	hours	days
cultivation time	days	weeks
products	extracellular	intracellular
media costs ($m^{-3})	6	50

[25] J.R. Mor, in 'Flavour Science and Technology', ed. Y. Bessière and A.F. Thomas, Wiley, Chichester, 1990, p. 135.
[26] A. Hatanaka, T. Kajiwara, and J. Sekiya, in 'Biogeneration of Aromas', ed. T.H. Parliment and R. Croteau, ACS Symposium Series No. 317, American Chemical Society, Washington DC, 1986, p. 167.

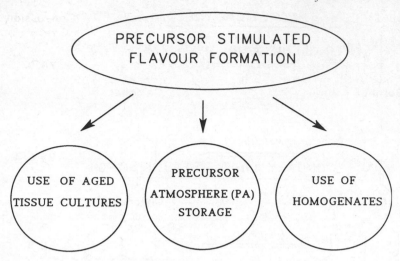

Figure 4 *Possibilities of precursor-stimulated flavour formation*

isomeric hexenals, as well as the corresponding C_6 alcohols from linoleic and linolenic acid respectively, by a sequence of enzymic reactions (acyl hydrolase — lipoxygenase — hydroperoxide lyase — isomerase — alcohol dehydrogenase), industrially interesting amounts of C_6 compounds, after incubation of fatty acid precursor, have not been achieved as yet. Literature data show a limit of (*E*)-hex-2-enal formation of about 350 mg kg^{-1} fresh weight for homogenates from quick grass (Agropyron repens) and apple peel after feeding with fatty acid precursor.[27] Apart from the low theoretical yield of C_6 aldehyde (Figure 5), the reasons for the limited formation of these industrially attractive flavour compounds have not been clarified. One of the factors that must be considered is secondary lipid metabolism, giving rise to transformation of enzymatically liberated fatty acids to new galactolipids that are not useful substrates for acyl hydrolase activity[28] (Figure 5). In spite of these limitations of C_6 aldehyde formation in homogenized plant tissues, a recent patent shows the practical application of this technique.[29]

The same concept, *i.e.* using homogenates, but from mushrooms[30] (Figure 6), has been recommended recently for bioflavour production. A process for the production of the characteristic mushroom flavouring, (*R*)-oct-1-en-3-ol, has been developed, which comprises homogenizing mushrooms (*Agaricus bisporus*; *A. bitorquis*; *A. campestris*) and, during or after homogenization, contacting the mushrooms with an aqueous medium containing a water-soluble salt of linoleic acid and oxygen.[31]

27 F. Drawert, A. Kler, and R.G. Berger, *Lebensm. Wiss. u. Technol.*, 1986, **19**, 426.
28 J. Joyard and R. Douce, in 'The Biochemistry of Plants', ed. P.K. Stumpf, Academic Press, Orlando, 1987, Vol. 9, p. 215.
29 S.K. Goers, P. Ghossi, J.T. Patterson, and C.L. Young, US P. 4 806 379, 1989.
30 M. Wurzenberger and W. Grosch, *Biochim. Biophys. Acta*, 1984, **795**, 163.
31 L.A. Kibler, Z. Kratky, and J.S. Tandy, Eur. P. Appl., 0 288 773, 1988.

1. <u>LOW THEORETICAL YIELD:</u>

LINOLENIC ACID: MW = 278
(<u>E</u>)-HEX-2-ENAL: MW = 98

THEORETICAL YIELD: 35 %.

2. <u>SECONDARY LIPID METABOLISM:</u>

TRANSFORMATION OF ENZYMATICALLY LIBERATED FATTY ACIDS
TO NEW GALACTOLIPIDS THAT ARE NOT USEFUL SUBSTRATES FOR
ACYL HYDROLASE ACTIVITY

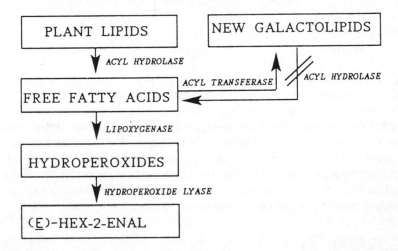

Figure 5 *Reasons for the limited yield of* C_6 *volatiles obtained in homogenized leaves*

2.2 Microbial Methodologies

A great number of reviews has been provided dealing with microbial and enzymic methods for the production of biologically important compounds.[32] There are two principal ways for the utilization of micro-organisms, *i.e.* (i) fermentation and (ii) microbial transformation. Fermentation has been practiced for the production of food since ancient times. Meanwhile, it has become an effective technology for the production of organic acids, antibiotics, amino acids, and derivatives of nucleic acids. In fermentation, cheap C- or N-sources are used, and the product is generated by the

---

[32] P. Winterhalter and P. Schreier, in 'Flavour Research', ed. R. Teranishi and T. Acree, ACS Professional Reference Ser., ACS, Washington DC, in press.

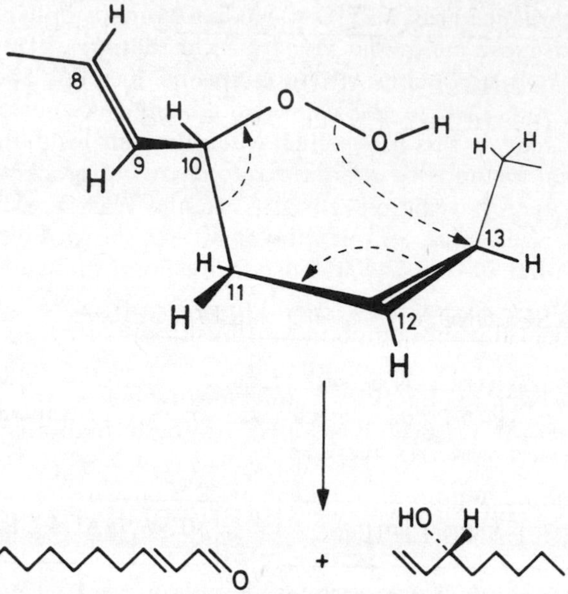

Figure 6 *Mechanism proposed to explain the action of hydroperoxide lyase on* 10-(S)-*hydroperoxyoctadeca*-trans-8-cis-12-*dienoic acid*[30]

complex metabolism of the micro-organism. Whereas fermentation requires C- and N-sources, a useful substrate is necessary for microbial conversion. The enzymic catalysis effects a simple and specific transformation of the substrate molecule. The substrate does not have to be 'natural'; 'non-natural' substrates can be biotransformed also. The advantages and disadvantages of both technologies are summarized in Table 2.

2.3 Fermentations

Fermentation, which was often an empirical use of micro-organisms, has been practised in the production of food since ancient times. Such classical

Table 2 *General properties of microbial transformations and fermentations*

	microbial transformation	*fermentation*
micro-organisms	growing, permanent, or treated cells	growing cells
reaction	simple (one- or multiple-step) catalytic reaction	complex reaction chain
reaction time	short	long
substrate	specific (sometimes expensive)	cheap C- and N-sources
product	natural or non-natural	natural
product concentration	high	low
product isolation	simple	laborious

fermentations lead to complex flavour compositions similar to those obtained in more controlled modern fermentations.[33] However, the potential of flavour-producing micro-organisms has not been systematically studied as yet. Due to the expected advantages, there is an increasing research activity in this field, which will ultimately result in the development of microbial methods as alternatives to chemical processes.[34]

Higher organized micro-organisms, such as yeasts and filamentous fungi, are able to synthesize complex flavours.[35] As shown by Figure 7, not only mushroom-like flavours,[36,37] but also fruity, floral, and spicy notes have been observed.[38]

The impact flavour compound in mushrooms is (*R*)-oct-1-en-3-ol. Early attempts to produce a flavour-rich mushroom protein using *Morchella* species go back to the 1950s[39] and, meanwhile, a new process with this micro-organism has been developed leading to high amounts of (*R*)-oct-1-en-3-ol.[40]

The impact compounds of blue cheese, *i.e.* alkan-2-ones, can be produced with strains of *Penicillium roquefortii* in submerged or surface culture in a short time.[41]

Esters are important aroma compounds formed by many micro-organisms. In the past, various yeasts and several deuteromycetes have been extensively investigated. Representative examples are *Geotrichum* species,[42] *Hansenula mrakii*,[43] and *Candida utilis*.[44] Successful results in ester formation have also been obtained using enzyme preparations of fungal origin, exhibiting lipase or lipase/esterase activity.

Another important attribute of whole microbial cells is for the production of optically active compounds. As early as the 1960s, the synthesis of optically active γ- and δ-lactones by microbial reduction has been demonstrated.[45] The fungus *Trichoderma reesei* showed the formation of 6-pentyl-α-pyrone,[46] and the yeast *Sporobolomyces odorus* can produce γ-decalactone

[33] W.P. Hammes, in 'Enzyme in der Lebensmitteltechnologie', 2. Symp. 1988, ed. K.H. Kroner, K. Lösche, and R.D. Schmid, GBF Monographien Bd. 11, VCH Verlagsgesellschaft, Weinheim, 1989, p. 3.
[34] D.H.G. Crout and M. Christen, in 'Modern Synthetic Methods 1989', ed. C.R. Scheffold, Springer, Berlin, Heidelberg, New York, 1989, p. 1.
[35] R.G. Berger, F. Drawert, and S. Hädrich, in 'Bioflavour '87', ed. P. Schreier, W. de Gruyter, Berlin, New York, 1988, p. 415.
[36] H.P. Hanssen, *Dtsch. Lebensm. Rdsch.*, 1982, **78**, 435.
[37] C.C. Chen and C.M. Wu, *J. Food Sci.*, 1984, **49**, 1208.
[38] H.P. Hanssen, *GIT Fachz. Labor.*, 1989, 996.
[39] J. Szuecs, US P. 2 850 841, 1958.
[40] Hüls A.G., US P. 4 810 504, 1989.
[41] J. Wink, U. Fricke, H.M. Deger, J. Mixich, D. Bauer, and U. Justinski, DE 3 701 836, 1988.
[42] A. Latrasse, P. Dameron, M. Hassani, and T. Staron, *Sci. Aliment.*, 1987, **7**, 637.
[43] L. Janssens, H.L. De Pooter, E.J. Vandamme, and N.M. Schamp, *Med. Fac. Landbouw. Rijksuniv. Gent*, 1987, **52**, 1907.
[44] D.W. Armstrong, S.M. Martin, and H. Yamazaki, *Biotechnol. Bioeng.*, 1984, **26**, 1038.
[45] .G. Tuynenburg Muys, B. van der Ven, and A.P. de Jonge US P. 3 076 750, 1963.
[46] R.P. Collins and A.F. Halim, *J. Agric. Food Chem.*, 1972, **20**, 437.

FLAVOUR COMPOUNDS FROM FUNGI			FLAVOUR COMPOUNDS FROM FUNGI		
STRUCTURE	ODOUR	FUNGI	STRUCTURE	ODOUR	FUNGI
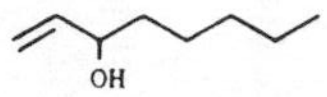	MUSHROOM-LIKE	NUMEROUS SPECIES	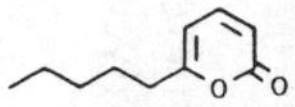	COCONUT-LIKE	TRICHODERMA spp.
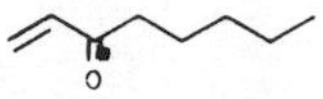	MUSHROOM-LIKE	NUMEROUS SPECIES	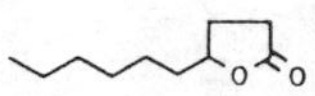	PEACH-LIKE	SEVERAL SPECIES
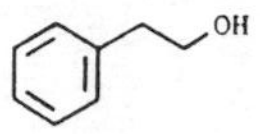	ROSE-LIKE	NUMEROUS SPECIES	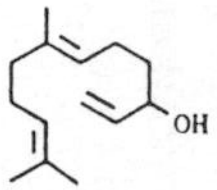	FLOWERY	SEVERAL ASCO- AND BASIDIOMYCETES
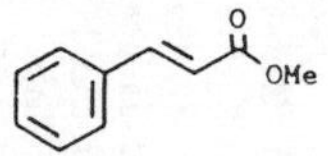	FRUITY, SWEET	SEVERAL BASIDIOMYCETES	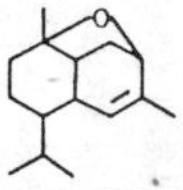	WOODY	LENTINUS LEPIDEUS
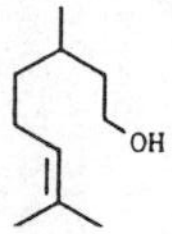	SWEET, ROSE-LIKE	NUMEROUS SPECIES	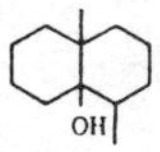	EARTHY	CHAETOMIMUM GLOBOSUM

Figure 7 *Flavour compounds from fungi* (*selected according to Hanssen*[38])

and *cis*-dodec-6-en-4-olide, responsible for a 'peach'-like odour impression.[47] Various *Candida* strains or fungi imperfecti are used on an industrial scale to produce γ-decalactone from castor oil and castor oil hydrolysate.[48-51] In addition, a number of different fungi, *i.e.* fungi imperfecti, ascomycetes, and basidiomycetes, forming lactones by *de novo* biosynthesis, have been detected (Table 3).[38, 46, 47, 52-58]

Table 3   De novo *biosynthesis of lactones by fungi*[38]

Fungus	*Sensory impression of culture*	*Reference*
deuteromycetes		
Fusarium poae	fruity, peach-like	52
Pityrosporum spp.	fruity, peach-like	53
Sporobolomyces odorus	peach-like	47
Trichoderma spp.	coconut-like	46
ascomycetes		
Monilia fruticola	peach-like	54, 55
basidiomycetes		
Bjerkandera adusta	sweet-aromatic, vanilla	56
Ischnoderma benzoinum	sweet	57
Polyporus durus	coconut, pineapple-like	58
Poria aurea	sweet, fruity	56
Tyromyces sambuceus	peach-, passionfruit-like	56

Since the first studies carried out with *Ceratocystis*,[59] the ability of fungi to produce terpenes—at least acyclic monoterpenes and several sesquiterpenes —has been generally acknowledged.[60] Meanwhile, several strains of *Ceratocystis* are among the most extensively studied fungi.[61] Whereas terpene

47 S. Tahara, K. Fujiwara, H. Ishizaka, J. Mizutani, and Y. Obata, *Agric. Biol. Chem.*, 1972, **36**, 2585.
48 M.I. Farbood and B.J. Willis, US P. 4 560 656, 1982.
49 P.S.J. Cheetham, K.A. Maume, and J.F.M. De Rooji, Eur. P. Appl., 0 258 993, 1987.
50 M.I. Farbood, J.A. Morris, M.A. Sprecher, L.J. Bienkowski, K.P. Miller, M.H. Vock, and M.L. Hagedorn, Eur. P. Appl., 0 354 000, 1989.
51 A.L.G. Boog, A.M. Van Grinsven, A.I.J. Peters, R. Roos, and A.J. Wieg, Eur. P. Appl., 0 371 568, 1989.
52 J. Sarris and A. Latrasse, *Agric. Biol. Chem.*, 1985, **49**, 3227.
53 J.N. Labows, K.J. McGinley, L.L. Leyden, and G.F. Webster, *Appl. Environ. Microbiol.*, 1979, **38**, 412.
54 J. Wink, A. Lotz, and P. Präve, *BTF Biotech. Forum*, 1987, **4**, 235.
55 J. Wink, H. Voelskow, S. Grabley, and H.M. Deger, Eur. P. Appl., 0 283 950, 1988.
56 R.G. Berger, K. Neuhäuser, and F. Drawert, *Flav. Fragr. J.*, 1986, **1**, 181.
57 R.G. Berger, K. Neuhäuser, and F. Drawert, *Biotechnol. Bioeng.*, 1987, **30**, 987.
58 F. Drawert, R.G. Berger, and K. Neuhäuser, *Chem. Mikrobiol. Technol. Lebensm.*, 1983, **8**, 91.
59 R.P. Collins and A.F. Halim, *Lloydia*, 1970, **33**, 481.
60 E. Sprecher and H.P. Hanssen, in 'Topics in Flavour Research', ed. R.G. Berger, S. Nitz, and P. Schreier, Eichhorn, Marzling, 1985, p. 387.
61 H-P. Hanssen, in 'Progress in Terpene Chemistry', ed. D. Joulain, Edition Frontieres, Gif-sur-Yvette, 1986, p. 331.

formation in yeasts is mainly restricted to a few compounds,[62] successful production of monoterpenes has been obtained recently using a *Saccharomyces cerevisiae* mutant.[63]

2.4 Biotransformations

Recently, a comprehensive review about the possibilities of microbial conversions in bio-organic chemistry has been provided.[34] In flavour chemistry, particular attention has been directed to the microbial bioconversion of terpenes. Principal types of reactions such as partial degradation, hydroxylation, epoxidation, hydration of double bonds, and reduction have been observed in a number of studies of terpene conversions using various species of micro-organisms.[32] They have also been found in our recent biotransformation study using a common fungus, well known in winemaking, namely *Botrytis cinerea*.[64] As shown in Figure 8, using α-damascone **1** as precursor, 3-oxo-α-damascone **3**, *cis*- and *trans*-3-hydroxy-α-damascone **4, 5**, γ-damascenone **6**, 3-oxo-8,9-dihydro-α-damascone **7** and *cis*- and *trans*-3-hydroxy-8,9-dihydro-α-damascone **8,9** have been identified as conversion products formed by *B. cinerea*. In addition, acid catalysed chemical transformation of **1** to the diastereomers of 9-hydroxy-8,9-dihydro-α-damascone **2A/2B** has been observed.[65] These biotransformation products are attractive due to their flavour activity and as flavour precursors. In principle, reactions of such type are examples of the conversion of easily available flavour chemicals to higher-valued products.

2.5 Enzymes

The International Union of Biochemistry has listed more than 2000 enzymes classified, from which more than 800 cell-free enzymes are commercially available. If a desired enzyme cannot be obtained from any of the major suppliers, a source, if it exists, can be identified through a suitable computer database for enzymes (ENZYDEX, BRENDA).[66, 67]

Many commercial procedures have been developed in enzyme technology, *e.g.* for the production of amino acids, derivatives of nucleic acids, sugars, and lipids. Some of them are exclusively prepared by the use of enzymes, *i.e.* L-alanine (with aspartate-β-decarboxylase), L-aspartic acid (with aspartase), high fructose containing starch syrup (with glucose isomerase), and 6-aminopenicillic acid (with penicillin acylase). A large number of chemicals applied for analyses in clinical chemistry, biochemistry, and pharmacology

[62] R. Hock, I. Benda, and P. Schreier, *Z. Lebensm. Unters.-Forsch.*, 1984, **179**, 450.
[63] F. Karst and B.D.V. Vladescu, Eur. P. Appl., 0 313 465, 1988.
[64] P. Brunerie, I. Benda, G. Bock, and P. Schreier, in 'Bioflavour '87', ed. P. Schreier, W. de Gruyter, Berlin, New York, 1988, p. 435.
[65] E. Schoch, I. Benda and P. Schreier, *Appl. Environ. Microbiol.*, 1991, **57**, 15.
[66] ENZYDEX, Enzyme Data Base, operated by Biocatalysts Ltd., Treforest Industrial Estate, Pontypridd, Mid Glamorgan, CF37 5UT, UK.
[67] D. Schomburg, *Nachr. Chem. Technik Labor.*, 1988, **36**, 374.

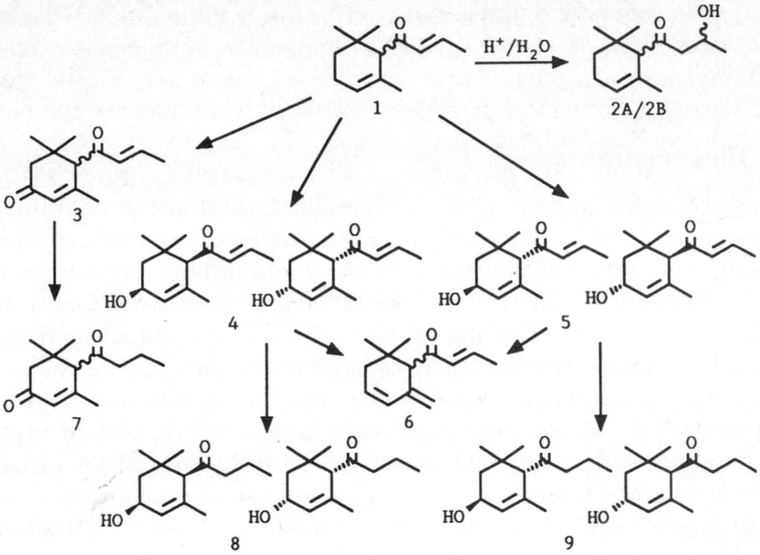

Figure 8 *Structures of bioconversion products formed from α-damascone* **1** *by* Botrytis cinerea[65]

is also produced by enzymes. Furthermore, several other procedures have already reached practical importance, *e.g.* the preparation of mevalono-lactone, antibiotics, pheromones, α-tocopherol, and muscone.

Using enzymes for synthetic purposes, it is most important to appreciate the major differences between these and classical chemical procedures, *i.e.*,

(i)   The catalysts are proteins or a protein and a low-molecular cofactor. They are relatively delicate systems and need to be handled with appropriate care.

(ii)   The reactions catalysed are reversible.

(iii)   Reactions occur in a narrow temperature (commonly between 20 and 40 °C) and pH range.

(iv)   The properties of the biocatalyst may vary with the form in which it is used (*e.g.* free or immobilized).

(v)   The enzyme may be susceptible to inhibition by the substrate(s), product(s), or components of the reaction medium.

(vi)   Some cofactors, in cofactor-dependent reactions, may be altered during a catalytic cycle. In such cases, the cofactor has to be recycled to the active form.

Much of the available information in this field has been collected in a number of reviews.[68-71]

[68] 'Enzymes in Organic Synthesis', ed. R. Porter and S. Clark, CIBA Foundation Symposium 111, Pitman, London, 1985.

A survey[34] about the actual use of enzymes shows that 65 % of all reported applications fall nearly equally into the classes of esterolytic reactions (40 %) and dehydrogenase reactions (25 %). However, applications of dehydrogenases are mostly performed by use of whole micro-organisms (*cf.* above) and not by isolated enzymes. Next in importance are oxygenase-mediated reactions, and peptide and oligosaccharide syntheses (together 24 %). Only a few reports about enzymatic carbon–carbon synthesis are available (2 %). All other reaction types comprise about 9 % of the total. Actually, just four biocatalytic systems predominate: pig liver esterase (PLE), pig pancreatic lipase (PPL), and the lipase from *Candida cylindracea* and from baker's yeast. This indicates the direction in which chemists looking to use enzymes are most likely to achieve success.

In discussing the potential use of enzymes for the production of flavour compounds, it has to be considered that in plant tissues, a certain amount of the flavour constituents are bound in non-volatile form, from which they can be liberated by the use of enzymes, *i.e.* glycosidases.[72, 73] In recent years, increasing research activity has been devoted to the study of conjugated forms of plant volatiles. The recent development of a continuous process of enzymic treatment of plant tissues to liberate aglycones (SECE: simultaneous enzyme catalysis extraction)[74] now opens the way for potential technological application, in which industrially attractive plant volatiles, not commonly available, will be accessible from their non-volatile conjugates. The system is generally suitable for enzymic reactions, in which apolar products are enzymatically formed from polar substrates.

A review about our actual knowledge of the conjugated forms of aroma compounds in fruits has been provided recently.[32] In addition, a high number of aglycones liberated by glycosidase activity has been identified in many fruit species, such as apples,[74] grapes,[75] papaya,[76] pineapple,[77] passion fruit,[78] several *Prunus* species,[79] sour cherry,[80] quince,[81] and raspberry.[82] In nearly all cases, the identified aglycones fell mainly into three categories

[69] C. Fuganti and S. Servi, in 'Bioflavour '87', ed. P. Schreier, W. de Gruyter, Berlin, New York, 1988, p. 555.
[70] M.P. Schneider and K. Laumen, in 'Bioflavour '87', ed. P. Schreier, W. de Gruyter, Berlin, New York, 1988, p. 483.
[71] 'Biocatalysts in Organic Media', ed. C. Lanne, J. Tramper, and M.D. Lilly, Elsevier, Amsterdam, 1987.
[72] P.J. Williams, C.R. Strauss, and B. Wilson, *J. Agric. Food Chem.*, 1980, **28**, 766.
[73] P.J. Williams, C.R. Strauss, and B. Wilson, *Phytochemistry*, 1980, **19**, 1137.
[74] W. Schwab and P. Schreier, *J. Agric. Food Chem.*, 1988, **36**, 1238.
[75] P.J. Williams, M.A. Sefton, and B. Wilson, in 'Flavour Chemistry: Trends and Developments', ed. R. Teranishi, R.G. Buttery, and F. Shahidi, ACS Symposium Series No. 388, American Chemical Society, Washington DC, 1989, p. 35.
[76] W. Schwab, C. Mahr, and P. Schreier, *J. Agric. Food Chem.*, 1989, **37**, 1009.
[77] P. Wu, M.C. Kuo, K.Q. Zhang, T.G. Hartmann, R.T. Rosen, and C.T. Ho, *Perfumer & Flavourist*, 1990, **15**, 51.
[78] P. Winterhalter, *J. Agric. Food Chem.*, 1990, **26**, 452.
[79] G. Krammer, P. Winterhalter, M. Schwab, and P. Schreier, *J. Agric. Food Chem.*, 1991, **39**, 778.
[80] W. Schwab, and P. Schreier, *Z. Lebensm. Unters.-Forsch.*, 1990, **190**, 228.
[81] P. Winterhalter and P. Schreier, *J. Agric. Food Chem.*, 1988, **36**, 1251.
[82] A. Pabst, D. Barron, P. Etievant, and P. Schreier, *J. Agric. Food Chem.*, 1991, **39**, 173.

biogenetically derived from (i) fatty acid, (ii) shikimate, and (iii) mono- and C_{13} norisoprenoid metabolism. In particular, the last-mentioned group has attracted the attention of researchers in flavour chemistry because of its important role in flavour genesis. The structures of six essential C_{13} norisoprenoid flavour precursors are outlined in Figure 9. The ketoalcohol **I** is known as synthetic precursor of megastigmatrienones.[83] The 3- and 4-hydroxylated β-ionol derivatives **II** and **III** play an important role in the formation of numerous C_{13} norisoprenoid flavourings.[84] Recently, compound **IV** has been recognized as a precursor of isomeric theaspiranes.[85] The triol **V** has been considered as progenitor of isomeric vitispiranes in grapes[86] and, finally, the acetylenic diol **VI** has been found recently as a precursor of damascenone.[87, 88]

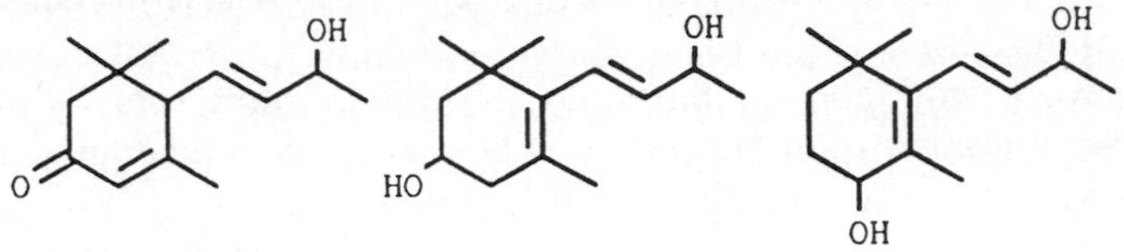

I | II | III

GRAPE PASSIONFRUIT | QUINCE PAPAYA | QUINCE PASSIONFRUIT

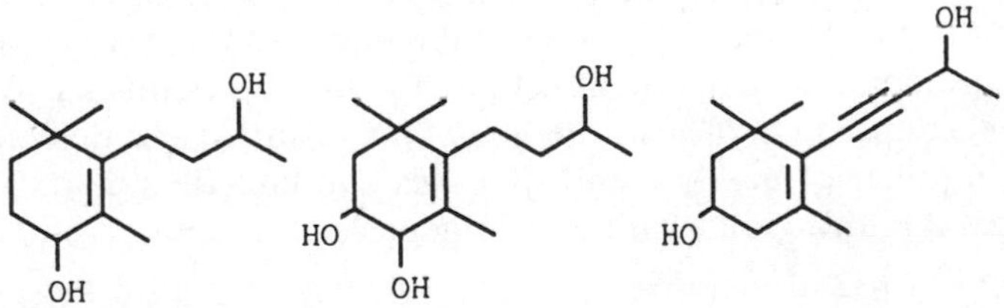

IV | V | VI

PASSIONFRUIT | GRAPE | GRAPE

Figure 9 *Structures and occurrence of essential* C_{13} *norisoprenoid flavour precursors.* **I** = *3-oxo-α-ionol;* **II** = *3-hydroxy-β-ionol;* **III** = *4-hydroxy-β-ionol;* **IV** = *4-hydroxy-7,8-dihydro-β-ionol;* **V** = *megastigm-5-ene-3,4,9-triol;* **VI** = *megastigm-5-ene-7-yne-3,9-diol*

[83] A.J. Aasen, B. Kimland, S.O. Almquist, and C.R. Enzell, *Acta Chem. Scand.*, 1972, **26B**, 2573.
[84] P. Winterhalter and P. Schreier, in 'Thermal Generation of Aromas', ed. T.H. Parliment, C.T. Ho, and R.J. McGorrin, ACS Symposium Series No. 409, American Chemical Society, Washington DC, 1989, p. 320.
[85] P. Winterhalter and P. Schreier, *J. Agric. Food Chem.*, 1988, **36**, 560.
[86] P. Winterhalter, M.A. Sefton, and P.J. Williams, *Am. J. Enol. Vitic.*, 1990, **41**, 227.
[87] M.A. Sefton, G.K. Skouroumounis, R.A. Massy-Westropp, and P.J. Williams, *Aust. J. Chem.*, 1989, **42**, 2077.
[88] P. Winterhalter, G. Full, M. Herderich, and P. Schreier, *Phytochem. Anal.*, 1991, **2**, 93.

Additionally, increasing industrial research activity is being devoted both to the isolation and separation of specific glycosidases from suitable sources,[89, 90] and the possibilities to use the still unknown potential of bound flavour constituents in plant tissues.[91]

2.6 Genetic Engineering

On the eve of the signing of the European directive on the commercial use of genetically modified organisms, it is worthwhile spending a few words on this methodology. Opportunities for genetic engineering in the flavour and food industry were recognized in the early 1980s. Recently, a survey has been provided about the various fields of flavour and food research, in which genetic engineering already plays a role or may be of importance in the future.[92] The areas considered comprise (i) products, (ii) processes, and (iii) raw materials.

As to flavour production, in particular, a number of microbially produced flavour compounds can be proposed as targets for genetic engineering, *e.g.* 5'-nucleotides, diacetyl, pyrazines, and lactones.

Concerning processes, for brewers' yeast various studies using genetic engineering have been published.[93–95] In malolactic fermentation, genetic engineering of yeast strains for winemaking has also been reported.[96]

Genetic engineering of plants and animals in order to adapt raw materials to more preferred specifications for the flavour industry is a long-term option. Interfering in the secondary metabolism of plants to manipulate flavours will require introduction or elimination of enzymes. In the case of plants it is a much more complicated situation than for micro-organisms. Whereas for the latter, elimination of just one gene can result in the desired effect, in plants a number of nearly identical genes can encode similar proteins. In such a 'gene family' some hundred of genes may occur.

3 Analytical Aspects

It is well established that chiral discrimination is an essential principle of odour perception.[97] The determination of the enantiomeric composition of aroma compounds is therefore of high importance in flavour analysis.[98] The

[89] O. Shoseyov, I. Chet, B.A. Bravdo, and R. Ikan, Eur. P. Appl., 0 307 071, 1988.
[90] Z. Gunata, S. Bitteur, R. Baumes, J.M. Brillouet, C. Tapiero, C. Bayonove, and R. Cordonnier, Eur. P. Appl., 0 332 281, 1989.
[91] C. Ambid, Eur. P. Appl., 0 309 339, 1988.
[92] N. Overbeeke, in 'Biotechnology-Challenges for the Flavour Industry', ed. R.C. Lindsay and B.J. Willis, Elsevier Applied Science, London, New York, 1989, p. 87.
[93] S. Holmberg, *Trends Biotechnol.*, 1984, **2**, 98.
[94] C.J. Panchal, I. Russell, A.M. Sills, and G.G. Stewart, *Food Technol.*, 1984, **38**, 99.
[95] A. von Wettstein, *Anth. v. Leeuvenhoek*, 1987, **53**, 299.
[96] R. Snow, *Food Technol.*, 1985, **39**, 96.
[97] G. Ohloff, 'Riechstoffe und Geruchssinn. Die molekulare Welt der Düfte', Springer, Berlin, Heidelberg, New York, 1990.
[98] A. Mosandl, *Food Rev. Intern.*, 1988, **4**, 1.

development of sensitive capillary gas chromatographic (HRGC) methods, for the enantio-differentiation of chiral flavour compounds occurring in natural matrices,[99, 100] has stimulated the interest in chirality evaluation of aroma substances exhibiting pronounced sensory properties, *e.g.* γ- and δ-lactones. Direct resolution of lactone enantiomers is the most elegant approach, and excellent results have been obtained using alkylated/acylated cyclodextrins as stationary phases.[101,102] The resolution of γ-lactone enantiomers has been achieved by use of hexakis(3-*O*-acetyl-2,6-di-*O*-pentyl)-α-cyclodextrin (Lipodex B)[103] and heptakis(3-*O*-acetyl-2,6-di-*O*-pentyl)-β-cyclodextrin (Lipodex D).[104] The determination of the enantiomeric composition of γ-lactones in complex natural matrices using multi-dimensional capillary gas chromatography (MDGC) and MDGC-mass spectrometry (MDGC-MS) has been described recently.[105-107] Using the same technique more recently, the direct chiral evaluation of δ-lactones in various fruit tissues by use of Lipodex D has been reported.[108]

Among the alkan-4-olides mentioned, γ-decalactone is widely used industrially in many fruit aroma compositions; in particular, in peach flavours. In addition to the synthetic ('nature-identical') product, γ-decalactone from microbial sources ('natural') has been introduced recently into the flavour industry market. In the control of flavoured food, methods of differentiation between the 'natural' and 'nature-identical' status of this industrially important flavour compound are of particular interest. These methods comprise the above mentioned chiro-specific MDGC and MDGC-MS, as well as on-line HRGC-isotope ratio mass spectrometric (HRGC-IRMS) analysis.[109] In Table 4, the enantiomeric composition of γ-decalactone from various natural sources is represented. In Table 5, the δ ^{13}C values measured for γ-decalactone originating from various sources are outlined.[110] As shown from Tables 4 and 5, both methods allow distinct differentiation between the 'natu-

99 R. Tressl and K.H. Engel, in 'Analysis of Volatiles', ed. P. Schreier, W. de Gruyter, Berlin, New York, 1984, p. 323.
100 R. Tressl, K.H. Engel, W. Albrecht, and H. Bille-Abdullah, in 'Characterization and Measurement of Flavours', ed. D.D. Bills and C. Mussinan, ACS Symposium Series No. 289, American Chemical Society, Washington DC, 1985, p. 43.
101 W.A. König, S. Lutz, C. Colberg, N. Schmidt, G. Wenz, E. von der Bey, A. Mosandl, C. Günther, and A. Kustermann, *J. High Resolut. Chromatogr. Chromatogr. Commun.*, 1988, **11**, 621.
102 H.P. Nowotny, D. Schmalzing, D. Wistuba, and V. Schurig, *J. High Resolut. Chromatogr. Chromatogr. Commun.*, 1989, **12**, 383.
103 A. Mosandl, U. Hener, U. Hagenauer-Hener, and A. Kustermann, *J. High Resolut. Chromatogr. Chromatogr. Commun.*, 1989, **12**, 532.
104 A. Mosandl, A. Kustermann, U. Hener, and U. Hagenauer-Hener, *Dtsch. Lebensm. Rdsch.*, 1989, **85**, 205.
105 A. Bernreuther, N. Christoph, and P. Schreier, *J. Chromatogr.*, 1989, **481**, 363.
106 S. Nitz, H. Kollmannsberger, and F. Drawert, *Chem. Mikrobiol. Technol. Lebensm.*, 1989, **12**, 75.
107 S. Nitz, H. Kollmannsberger, and F. Drawert, *Chem. Mikrobiol. Technol. Lebensm.*, 1989, **12**, 105.
108 A. Bernreuther, J. Bank, G. Krammer, and P. Schreier, *Phytochem. Anal.*, 1991, **2**, 43.
109 H.L. Schmidt, *Fresenius Z. Anal. Chem.*, 1986, **324**, 760.
110 A. Bernreuther, J. Koziet, P. Brunerie, G. Krammer, N. Christoph, and P. Schreier, *Z. Lebensm. Unters.-Forsch.*, 1990, **191**, 299.

Table 4 *Enantiomeric composition of γ-decalactone from various natural sources*[110]

Source	(R) (%)	(S) (%)	e.e. (R) (%)
Apricot (Germany)[a]	94.3	5.7	88.6
Blueberry (Scandinavia)[a, b]	95.0	5.0	90.0
Guava (Brazil)	95.1	4.9	90.2
Mango (Venezuela)	65.8	34.2	31.6
(Israel)	68.3	31.7	36.6
(Mexico)	90.8	9.2	81.6
Passion Fruit			
yellow (Colombia)	91.3	8.7	82.6
red (Australia)	91.6	8.4	83.2
Peach (Germany)	87.4	12.6	74.8
(Italy)	86.3	13.7	72.6
(Greece)	89.0	11.0	78.0
Pineapple (Honduras)	83.3	16.7	66.6
Plum (blue, Germany)	93.1	6.9	86.2
(green, Germany, *i.e.* Reineclaude[a])	89.8	10.2	79.6
Reineclaude (green, Italy)	92.2	7.8	84.4
Reineclaude (yellow, Germany)	90.8	9.2	81.6
Mirabelle[a]	87.8	12.2	75.6
Mirabelle	86.2	13.8	72.4
Strawberry *(Fragaria ananassa)*			
Korona, Germany	100	0	100
Bel Ruby (Italy)	98.4	1.6	96.8
Senga Sengana (Poland)[a]	97.9	2.1	95.8
Cora (Spain)	100	0	100
Spadeka (France)	98.5	1.5	96.0
Bogota (France)	97.8	2.2	95.6
Fragaria vesca (Spain)	97.5	2.5	95.0
Microbial (A)	98.9	1.1	97.8
Microbial (B)	99.5	0.5	99.0

[a] Frozen fruits,
[b] In additional studies of six samples originating from other countries, γ-decalactone was not detectable (detection limit $< 1\mu g\ kg^{-1}$).

ral' and synthetic product. Moreover, in several cases, distinction between the natural origin is possible.

4   Conclusions

From the discussion of the principal methodologies for the biotechnological production of flavour compounds, the following conclusions can be drawn:

Table 5 ^{13}C-*Content of γ-decalactone originating from various sources*[110]

Origin	δ ^{13}C	‰	(PDB)
	HRGC-IRMS		Conventional
Natural			
Strawberry			
Spadeka	−28.2		
Bogota	−28.9		
Bel Ruby	−30.5		
Peach			
Italy	−40.9		
Germany	−38.5		
Plum (Mirabelle)	−39.6		
Apricot[a]	−38.0		
Microbial (A)	−31.2		−30.8
Microbial (B)	−30.3		−30.7
Synthetic			
Aldrich	−26.9		−27.3
Roth	−24.4		
Takasago	−26.0		

[a] Except for the value given for apricot that showed a variation of ± 2.0 ‰ due to interfering peaks; for the other HRGC-IRMS determinations, variations of ± 0.5 ‰ were found. For the conventional measurements ± 0.2 ‰ was determined.

1. The potential of plant cell cultures to produce specific secondary metabolites is of considerable interest in connection with their biotechnological utilization. However, the formation and accumulation of secondary metabolites do not occur in the cell cultures of higher plants. Nevertheless, plant cell cultures are considered to be useful for transforming cheap and plentiful substances into rare and expensive compounds.

2. Microbial biotransformation is a favourable methodology for converting inexpensive, easily available flavour chemicals—or their precursors—into higher-valued products. Microbial reactions are already used in flavour production, and both the technologies of fermentation and biotransformation will be more significant factors in the production of flavour chemicals in the future.

3. As to enzymes, the use of biocatalysts already plays a role in the stereoselective production of fine chemicals. This methodology will also open new ways for the production of flavour substances. Moreover, technologies will be developed to use the still unexploited potential of glycosidically bound aroma compounds by the application of hydrolases.

4. Genetic engineering has provided the first remarkable results; in partic-

ular, in scientific aspects. It has generated opportunities that will be topics of future research activities.

5.   As to analytical aspects, both the techniques of chiro-specific HRGC as well as on-line HRGC-IRMS will be of increasing importance for the determination of the origin of a commercial flavour substance. Whereas MDGC and MDGC-MS analyses of enantiomers are now fully established, the IRMS technique shows many weak-points. However, these problems should be solved by the future introduction of on-line HRGC-IRMS determination of hydrogen/deuterium ratios.

Acknowledgements

I wish to express my deep appreciation to my co-workers, who are identified in the references. Current studies are gratefully supported by the Deutsche Forschungsgemeinschaft, Bonn.

Naturally-occurring Flavours from Fungi, Yeast, and Bacteria

R.G. Berger

INSTITUT FÜR LEBENSMITTELCHEMIE DER UNIVERSITÄT, HANNOVER, WUNSTORFERSTR. 14, D-3000 HANNOVER 91, GERMANY

F. Drawert, and P. Tiefel

INSTITUT FÜR LEBENSMITTELTECHNOLOGIE UND ANALYTISCHE CHEMIE DER TU MÜNCHEN, D-8050 FRIESING 12, GERMANY

1 Introduction

Non-volatile flavour compounds such as organic acids, amino acids or 5'-nucleotides are biotechnologically produced on an industrial scale. Much less is known, however, about the genesis and biotechnology of volatile microbial flavours. In the beginnings of modern microbiology the odour characteristics of micro-organisms were used for taxonomic and diagnostic purposes, but their chemical basis remains largely unknown. A standard monograph on 'Fungal Metabolites', published in 1983, listed numerous non-volatile secondary compounds but devoted less than two out of the 560 pages to volatiles.[1]

This situation changed in the early eighties as the rapidly growing industrial demand for volatile flavours could no longer be satisfied from traditional plant sources. In Germany, one company admits that about 50% of their food flavours labelled 'natural' contain microbial volatiles; without them the sensory impression would remain non-competitive.

This chapter will highlight some selected recent work in this area rather than aim at an exhaustive overview of the immense number of papers recently published on this topic.

[1] 'Fungal Metabolites', ed. W.B. Turner and D.C. Aldridge, Academic Press, London–New York, 1983.

2 Bacteria

2.1 Food-related Micro-organisms

Procaryotic members of different genera play a key role in the manufacture
of classical fermented foods made from milk, flour, or meat. It is sometimes
difficult to assess whether the typical volatiles result directly from microbial
de novo synthesis or from the chemical degradation of precursors delivered
by microbial action. The dominating use of complex food substrates or food
imitating media in experimental systems, though evident from a practical
point of view, does not facilitate the discussion.

A tomato juice broth was used to investigate the volatile metabolites of 18
strains of lactic acid bacteria, representing the genera *Lactobacillus,
Pediococcus*, and *Leuconostoc*.[2] A total of 20 compounds (some of which are
shown in Figure 1) were identified and found suitable to differentiate
between the various organisms. Acetoin was among the volatiles, but

Mean concentration (μg l^{-1})

3-methyl-2-butanol	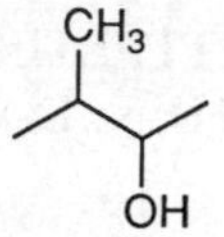 25.5 ± 0.4
ethyl 2-hydroxypropanoate	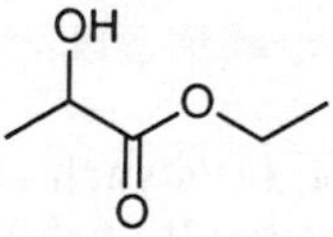 1122.1 ± 24.8
benzaldehyde	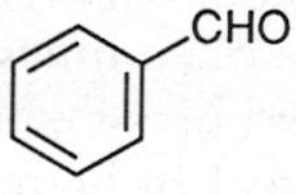 88.5 ± 1.1
3-methylthio-1-propanol	 233.8 ± 1.6
α, α-dimethylbenzylalcohol	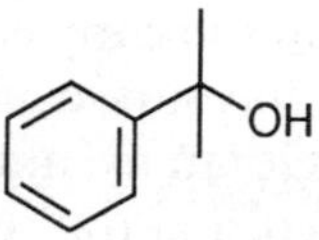 86.4 ± 0.8
benzylalcohol	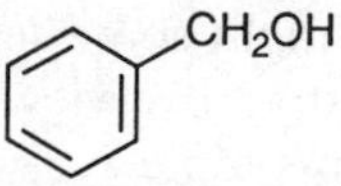 233.2 ± 0.8

Figure 1 *Volatile compounds of lactic acid bacteria*[2]

[2] R.P. Tracey and T.J. Britz, *Appl. Environ. Microbiol.*, 1989, **55**, 1617.

diacetyl was not reported to be present in the freon extracts. Most of the compounds identified showed chemically simple structures: α,α-dimethyl benzlyalcohol may have been derived from one of the constituents of the nutrient medium.

When sterilized milk was incubated with individual strains or mixtures of *Lactobacillus, Bifidobacterium,* and *Streptococcus,* the products obtained with the mixed starter culture yielded the highest sensory scores.[3] In the single starter culture controls, *Bifidobacterium breve* produced the highest acetaldehyde concentrations, while *B. longum* yielded the highest ethanol levels. Diacetyl was detected in the headspace of all samples. A Japanese patent[4] described the manufacture of diacetyl from citrate using *Streptococcus cremoris* and *S. diacetylactis.* Oxidizing agents were recommended before or after distillation of the preparation, underscoring the precursor role of acetoin and an unfavourable redox potential of the untreated nutrient broth. After the chemical oxidation step a remarkable product concentration of 14 g l^{-1} was reached.

Wine-related media inoculated with *Leuconostoc oenos* were investigated for headspace volatiles. Some aliphatic alcohols were the major constituents.[5] The benefits of malolactic fermentation are not disputable as far as non-volatile flavours are concerned, however, the contribution of *Leuconostoc* volatiles—based on the cited study—appears to be minor. By contrast, caproic acid bacteria imparted a distinct fruity note to qujiu, a Chinese alcoholic beverage.[6] Growing on inorganic carriers the bacteria produced caproic acid in abundance and this was subsequently transformed to its ethyl ester. Ester concentrations reached up to 4 g l^{-1} in the final beverage.

A series of papers from a Danish group[7] reported on the contribution of starter cultures to the flavour of sourdough bread (Figure 2) as determined by headspace/extraction GC-MS and sensory analysis. The sourdough rye bread crumbs generally had the highest contents of 2-propanone, 3-methyl butanal, benzylalcohol, and 2-phenylalcohol, while the chemically acidified doughs yielded crumbs that had high contents of ethyl lactate and hexanal. The sourdough derived products were found to possess a more intense and 'rye bread-like' odour than the chemically acidified samples. Consequently, further work considered aspects of the type of starter culture, fermentation temperature, and flour-water ratios.

2.2 Microbial Off-flavours

It is not the chemical structure of a volatile flavour compound but the chemical surroundings in which it occurs that determines if humans like or dislike

[3] H. Yuguchi, A. Hiramatsu, K. Doi, C. Ida, and S. Okonogi, *Food Feed Chem.,* 1989, **60**, 734.
[4] Y. Shukunobe and S. Takato, Jap. P. 01 168 256, 1989.
[5] M. le Roux, H.J.J. van Vuuren, L.M.T. Dicks, and M.A. Loos, *System. Appl. Microbiol.,* 1989, **11**, 176.
[6] X. Wang, Z. Liu, and X. Guo, Chin.N. 1 033 644, 1989.
[7] A. Hansen, B. Lund, and M.J. Lewis, *Lebensm. Wiss. Technol.,* 1989, **22**, 141.

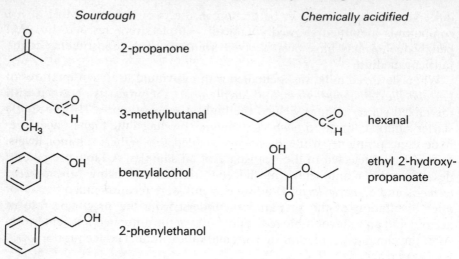

Figure 2 *Major aroma compounds in rye bread crumbs[7]*

the sensory results. The ability of micro-organisms to degrade natural polymers and monomers to volatile products may, hence, yield desired odours or off-odours. Many papers have dealt with microbial off-flavours in seafood, meat, and, more recently, in cereals such as barley or wheat.[8] In the course of spoilage of chicken carcasses,[9] for example, organoleptically detectable changes were paralleled by the formation of thio-compounds and free fatty acids, followed by the formation of fatty acid esters at the expense of corresponding alcohols and aldehydes. The microbial flora, largely consisting of *Pseudomonas*, produced a total of 86 volatile compounds. The authors' statement that microbial volatiles should not be used as indicators of spoilage, but rather the loss of primary volatiles should be taken as an indicator of freshness instead, is in contrast to most other preceding papers, but should be more accepted in the future.

Whilst discussing microbial off-flavours one should also point out the microbial ability to remove off-flavours, *e.g.* by reducing hexanal to hexanol in soy milk or by removing certain off-flavours from alcoholic beverages.[10] A recurrent problem in drinking water supplies is the formation of earthy-musty tastes and odours. Biodegradation of one of the 'bioflavours' responsible (2-methylisoborneol, a product of some actinomycetes and cyanobacteria) may offer an efficient means of removal. This compound is not oxidized by conventional water treatment methods, but was degraded by a mixed culture of aquatic *Pseudomonas* strains.[11]

[8] E. Wasowitz, E. Kaminski, H. Kollmannsberger, S. Nitz, R.G. Berger, and F. Drawert, *Chem. Mikrobiol. Technol. Lebensm.*, 1988, **11**, 161.
[9] S.H. Viehweg, R.E. Schmitt, and W. Schmidt-Lorenz, *Lebensm. Wiss. Technol.*, 1989, **22**, 346.
[10] R.G. Berger, F. Drawert, and S. Hädrich, in 'Bioflavour '87', ed. P. Schreier, W. de Gruyter, Berlin, 1988, 415.
[11] G. Izaguirre, R.L. Wolfe, and E.G. Means, *Appl. Environ. Microbiol.*, 1988, **54**, 2424.

2.3 Streptomycetes

Knowledge of the metabolism and specific properties of *Streptomycetes* has been significantly improved by the industrial interest in highly-active anti-biotics. The potential of these mycelium-like growing bacteria to form more complex volatile compounds has been barely considered. A German patent claimed some antibacterial lactones and furans.[12] Earlier work by Gerber[13] revealed the formation of terpenic alcohols and a substituted pyrazine. Volatile products such as geosmin are distinguished by ultra-low odour detection thresholds, and traces of microbially produced geosmin may cause problems in drinking water. On the other hand, traces of the same compound may act positively by rounding-off food flavours (such as beetroot or whisky) and also perfumes. A consideration of the synthesis of geosmin demonstrates the difficulties of building up the three chiral carbons chemically and the advantage of a biotechnological process is obvious. The crucial drawback of such an approach is the insufficient knowledge of the biochemical routes and enzymes involved. This is also true for most other complex microbial volatiles. A possible biosynthesis of methylisoborneol was proposed earlier,[13] but experiments supporting, or otherwise, the previous suggestion have not yet been published.

2.4 Biotransformation/Bioconversion

If a certain volatile flavour has been selected as an industrial target compound the most straightforward way may be to screen for micro-organisms converting a related natural starting material in a predictable manner. The conversion of pinene,[14] or, one step further, of pinene epoxide,[15] may serve as an example: *Pseudomonas* strains, or cell-free extracts derived thereof, oxidatively opened the C—C and C—O bridges yielding open chain terpenic aldehydes or alcohols. These products were claimed to be useful in perfumery and food formulations. An unusual degradation of cinnamic acid was reported for another strain of *Pseudomonas* resulting in the accumulation of acetophenone, a constituent of honey, plum, and strawberry flavour.[16] The reaction sequence starts with the addition of water to the double bond in the side chain, followed by dehydrogenation and decarboxylation of the intermediate 3-keto acid. Enterobacteriaceae of the genus *Erwinia*, well-known plant-pathogens, decarboxylated and deaminated naturally-occurring amino acids, thus yielding the corresponding alcohols.[17] In a second step the alcohols were converted to acetate esters. The resulting volatiles impart strong rose, potato, or floral aromas to food.

[12] S. Grabley, J. Wink, K. Kühlein, G. Seibert, K. Hütter, H. Uhr, and A. Zeeck, Eur. P. 337 152, 1989.
[13] N.N. Gerber, *Crit. Rev. Microbiol.*, 1979, 101.
[14] P.C. Harries, R. Jeffcoat, E.T. Griffith, and P.W. Trudgill, Eur. P. 271 609, 1988.
[15] K.J. Davis, D.J. Best, A.G. Burfield, F. Taylor, and J.S. Talbot, WO 8 911 536, 1989.
[16] M.D. Hilton and W.J. Cain, *Appl. Environ. Microbiol.*, 1990, **56**, 623.
[17] H.E. Spinnler and A. Djian, 2. Int. Symp. Biochem. Engin., Stuttgart, Germany, 1990.

3   Yeasts

3.1   Simulated Food Systems

Volatiles of alcoholic beverages originating from the metabolism of yeasts include simple aliphatic and aromatic alcohols, fatty acid esters thereof, carbonyls, lactones, thio-compounds, and some phenolics.[18] As early as in 1984 Schreier[19] cultured 18 yeast species and strains in a model system to show that terpene hydrocarbons and alcohols were not produced or produced in trace amounts only. It was concluded that the varietal classification of white wines based on terpene fingerprint patterns of grape cultivars is not affected by yeast metabolism. Confirming these results, an Italian group looked at the volatiles of musts of six different yeast strains.[20] Only seven out of 38 volatiles examined were highly controlled by the yeast strain, among them 2-phenylethyl acetate, ethyl butanoate, ethyl 4-hydroxybutanoate, and n-hexanol.

A different situation was reported for the fermentation of beer wort.[21] When brewers' yeast was cultured on synthetic glucose or maltose media it was found that the fusel oils and volatile acids formed on 10% sugar media were very similar to the composition of a bright lager beer. Obviously the minor constituents of wort do not serve as precursors in the yeast's metabolism of volatiles.

The production of wines and beers by flash fermentation technologies will clearly affect the patterns of volatiles. A US patent[22] described a method for accelerated fermentation using alcohol-tolerant and flavour-producing strains of *S. cerevisiae*, both immobilized on inert organic or glass beads. A mixed culture including *Lactobacilli* was suggested for flavour improvement. Reduced pressure during fermentation depressed the growth of yeast cells and the concentrations of acetate esters,[23] underscoring the primary character of the pathways involved.

The ability of many yeast strains to form high amounts of lactones has meanwhile found active industrial application. Amino acids may represent intermediate precursors of lactones as was suggested by a study using *S. fermentati* and a simulated sherry medium.[24] The addition of 1-[2-^{14}C]-glutamic acid yielded labelled γ-butyrolactone and further lactones and esters that are easily formally derived from the labelled amino acid.

[18] 'Volatile Compounds in Food', 5th Edn. ed. S van Straten and H. Maarse, TNO Zeist, The Netherlands, 1983.
[19] R. Hock, I. Benda, and P. Schreier, *Z. Lebensm. Unters. Forsch.*, 1984, **179**, 450.
[20] A. Cavazza, G. Versini, A. Dalla Serra, and F. Romano, *Yeast*, 1989, **5**, 163.
[21] K. Steiner, *Brau. Rundsch.*, 1989, **100**, 299.
[22] M.K. Hamdy, US P. 4 929 452, 1990.
[23] M. Sakaguchi, T. Hirose, K. Nakatani, M. Ohnishi, and J. Kumada, *Ferm. Bioind. Chem.*, 1990, **68**, 261.
[24] R.E. Wurz, R.E. Kepner, and A.D. Webb, *Am. J. Enol. Vitic.*, 1988, **39**, 234.

3.2 Non-food Systems

Yeast strains, and in particular baker's yeast, have a long tradition as catalysts in the non-food systems of the organic chemist.[25] Lactone formation,
for example, has been achieved by transforming a synthetic acetal acid to 3-
methyl-γ-butyrolactone.[26] The multi-step conversion included two stereoselective reductions yielding a product with a 90 % enantiomeric excess. The
fruity/floral aromas produced by yeasts grown on synthetic nutrient media
may usually be traced back to the formation of simple alcohols and to ethyl,
2-methylpropyl, and acetate esters;[27, 28] for example, the quince-like smell of
a mutant strain of *Geotrichum candidum*, formerly *Oospora lactis*, was due to
ethyl 2-methylpropanoate.[28] As some of these yeast flavours may gain industrial importance, attempts were made to increase the productivity of the cultures.

3.3 Improvement of Yields

When a microbial cell is placed at an increased osmotic pressure, the
intracellular equilibrium pressure has to be adjusted by either transport or
synthesis of metabolites. The effect of water activity on lactone production
by the yeast *Sporidiobolus salmonicolor* was investigated.[29] Decreased a_W values of 0.97 to 0.99 lead to increased aroma accumulation. Future work will
be required to show if the osmoregulatory process or an indirect stress effect
are able to induce synthesizing key enzymes.

Precursor feeding is a proven means of enhancing the yield of
biotechnological processes. A *Hansenula* yeast that was producing considerable amounts of esters was also capable of converting added alcohols into
corresponding acetates.[30] Fusel oil, a cheap fermentation by-product, was
rapidly converted to the corresponding acetates when added to the growing
culture (Figure 3). The volatile products were recovered from the synthetic
medium using gas stripping and adsorption on an activated carbon trap.

Fusel alcohol	Ester	Threshold in ng (Sniffing-GC)	Conversion (%)
2-methylbutanol	2-methylbutyl acetate	18–35	63
3-methylbutanol	3-methylbutyl acetate	2.9–4.0	91
2-methylpropanol	2-methylpropyl acetate	89–179	15

Figure 3 *Bioconversion of fusel alcohols to acetates by* Hansenula *yeast*[30]

[25] F. Drawert, H. Barton, and J. Beier, in 'Houben Weyl-Meth. Org. Chem', Vol.6/1d, ed. H. Kropf and G. Hesse, Thieme, Stuttgart, 1978, p. 299.
[26] P. Ferraboschi, P. Grisenti, A. Fiecchi, and E. Santaniello, *Org. Prep. Proced. Int.*, 1989, **21**, 371.
[27] B.H. Ahn, H.S. Kang, and H.K. Shin, *Food Feed Chem.*, 1988, **20**, 718.
[28] A. Latrasse, P. Dameron, M. Hassani, and T. Staron, *Sci. Aliments*, 1985, **7**, 637.
[29] P. Gervais and G. Battut, *Appl. Environ. Microbiol.*, 1989, **55**, 2939.
[30] L. Janssens, H.L. de Pooter, L. Demey, E.J. Vandamme, and N.M. Schamp, *Meded. Fac. Landbouwwet. Rijksuniv. Gent*, 1989, **54**, 1387.

It is increasingly recognized that an immediate removal of products from their producer cells can be the critical factor in terms of reaching high productivities. C_2 to C_7 alcohols were oxidized by the methylotrophic yeast *Pichia pastoris* at elevated pO_2 in a fed-batch process.[31] An amine buffer such as TRIS was added to trap chemically the aldehydes formed via Schiff bases. Thus, ethanol was converted into acetaldehyde at rates of up to 4 g l^{-1} h^{-1}. Physical trapping on solid polymers was successfully demonstrated using an ester-producing *Kluyveromyces marxianus* yeast, a monoterpene-producing *Ambrosiozyma monospora* yeast,[32] and using degraded brewer's yeast.[33] Natural or synthetic organic or inorganic polymers were examined; some polystyrene resins showed superior applicability in various biosystems.

4 Fungi

4.1 Mushroom-like Flavours

The mushroom-like flavour of edible fungi is mainly based on a group of volatile C_8-compounds (Figure 4). The subjects of earlier studies have been the identification of the structures, the elucidation of stereochemistry and sensory properties, and their biosynthesis from unsaturated C_{18} fatty acids via a lipoxygenase system.[34] The more abundant 1-octen-3-ol and the more

Compound	Structure	Threshold value (mg kg^{-1})	Flavour note
1-octen-3-ol	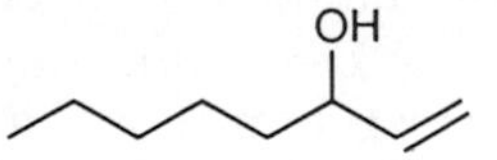	0.01	mushroom-like
1-octen-3-one	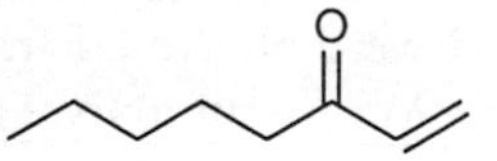	0.004	boiled mushroom, metallic at higher conc.
E-2-octen-1-ol		0.04	medicinal or oily
E-2-octenal	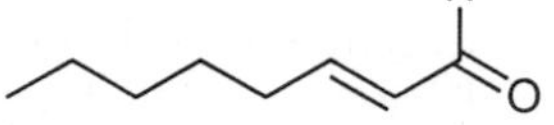	0.003	sweet-phenolic
3-octanol	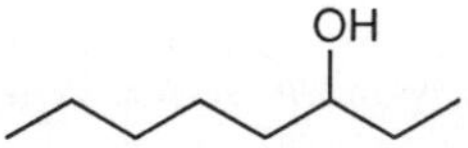	0.018	like cod liver oil

Figure 4 *Threshold values and odour characteristics of main volatiles of fresh mushroom*[34]

[31] W.D. Murray, S.J.B. Duff, and P.H. Lanthier, US P. 4 871 699, 1989.
[32] A. Klingenberg and H.P. Hanssen, *Chem. Biochem. Eng. Q.*, 1988, **2**, 222.
[33] T. Nakazawa, K. Usui, O. Koichi, and M. Ogawa, Jap. P. 63 102 650, 1988.
[34] J. Kinderlerer, *J. Appl. Bacteriol. Symp. Suppl.*, 1989, 133S.

odour-active corresponding ketone are thought to be the character-impact components of champignon and other edible species. An industrial process based on *Morchella* was described in which submerged cultured mycelia were accumulated in a batch process, and the stationary phase cells then subjected to high shear stress in the presence of linoleic acid.[35] The intimate contact of enzymes and substrate resulted in the formation of about 2.5 g octenol kg⁻¹ mycelial dry matter. The lipoxygenase-catalysed degradation of unsaturated fatty acids was also found in other genera of fungi. The mushroom-like odour attributes, *e.g.* of Camembert-type cheeses, depend on the same volatiles that are produced by *Penicillium* strains. The formation of these volatiles by *Penicillium* appears to occur on very different substrates in a similar manner.[36]

4.2 Other Volatile Compounds

During the past few years it has become evident that lower fungi (*Phycomycetes*) such as *Mucor*, and also representatives of the higher fungi (*Deuteromycetes, Ascomycetes, Basidiomycetes*) may be potent producers of more complex, industrially sought-after volatile flavours. *Aspergillus niger* strains transformed coconut fat to methylketones in a solid state fermentation,[37] and ionones to degradation products with tobacco-like odours.[38] In the latter case reactions included hydroxylation of the cyclohexene ring followed by oxidation of the alcoholic moiety or elimination of H_2O. A French group screened 29 lignolytic *Basidiomycete* strains on six different media and succeeded in identifying more than 100 volatile metabolites, some of them with unusual structures.[39] The volatile pattern of these organisms is often dominated by compounds possessing aromatic skeletons, thus reflecting the 'natural' chemical environment. One example concerns the vanillin-related methoxybenzaldehydes which are synthesized by *Ischnoderma benzoinum* in phenylalanine-supplemented media.[40] Mono- and sesquiterpenes frequently occur in submerged cultured strains of *Poria*.[10, 41] The chemical and physical environment of the fungus may significantly affect the odour profiles. Even more important may be strain specificity. Three strains of *Phlebia radiata*, another lignolytic *Basidiomycete*, were investigated for odorous volatiles, and α-bisabolol was the most abundant compound in two productive strains, whereas the third one did not form any volatile flavours.[42] Also

35 F. Schindler and R. Seipenbusch, *Food Biotechnol.*, 1990, **4**, 77.
36 C. Karahadian, J.B. Josephson, and R.C. Lindsay, *J. Agric. Food Chem.*, 1985, **33**, 339.
37 M. Humphrey, S. Pearce, and B. Skill, in 'Bioflavour '90', ed. J.R. Piggott and A. Paterson, Elsevier, Amsterdam, 1991, in press.
38 Y. Yoshiwara, H. Jumiko, A. Masatoshi, H. Tadaharu, and M. Yoichi, *Appl. Environ. Microbiol.*, 1988, **54**, 2354.
39 A. Gallois, B. Gross, D. Langlois, H.E. Spinnler, and P. Brunerie, *Mycol. Res.*, 1990, **94**, 494.
40 R.G. Berger, K. Neuhäuser, and F. Drawert, *Biotechnol. Bioengin.*, 1987, **30**, 987.
41 R.G. Berger, S. Hädrich-Meyer, and F. Drawert, in 'Proc. 11th Int. Congr. Ess. Oils, Frag. & Flav.', ed. S.C. Bhattacharyya, N. Sen, and K.L. Sethi, Vol. III, Oxford IBH Publ., New Dehli, 1989, p. 127.
42 B. Gross, A. Gallois, H.E. Spinnler, and D. Langlois, *J. Biotechnol.*, 1989, **10**, 303.

odourless was the CBS strain of *Poria subacida*, while another strain of the same species delivered a broad spectrum of volatile metabolites[39] (Figure 5). There are indications of genetic instability of some of the flavour yielding metabolic traits that could adversely affect larger-scale industrial applications. Prolonged refrigerator storage of the sub-cultures or improper nutrient conditions are suggested reasons for this instability.

Structure		Flavour	Concentration (μg l^{-1})
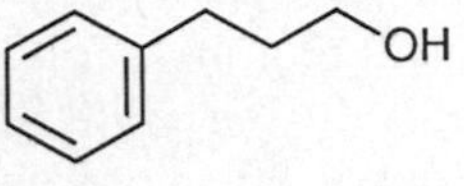	phenylpropanol	floral	50– 250
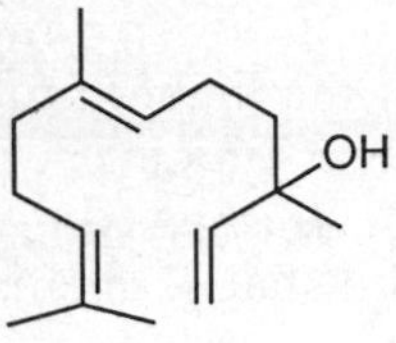	E-nerolidol	woody, floral warm	50– 250
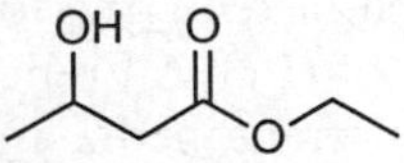	ethylhydroxy-butanoate	pineapple	250–1000
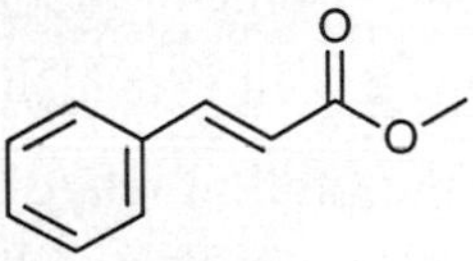	methylcinnamate	strawberry-like	50– 250
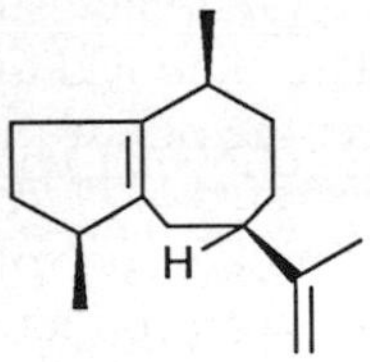	α-guaiene	balsamic	0– 50
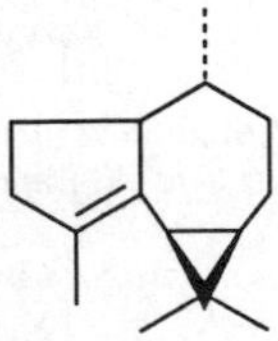	α-gurjunene	fatty	0– 50

Figure 5 *Volatile flavour compounds identified in liquid cultures of lignolytic basidiomycetes*[39]

4.3 Lactones

The biogeneration of volatile lactones has become an industrial success story. Yeasts or imperfect fungi such as *Fusarium poae*[43] or *Aspergillus*[44] were supposed to perform better on the industrial scale than higher fungi. Maximum yields of about 1 g l^{-1} were reported for the conversion of castor oil to γ-decalactone by a strain of *Aspergillus niger*.[44] Biotechnologically more demanding *Basidiomycetes* such as *Polyporus durus* or *Tyromyces sambuceus* reached similar concentration ranges for γ-octalactone and γ-decalactone, respectively, after an optimization protocol was conducted on the laboratory scale.[10] Semi-continuous cultures of *Tyromyces* that had been grown for more than four months reached peak concentrations of more than 1.5 g γ-decalactone l^{-1}.[45] Lactone formation appears to be the result of a metabolic overflow: when the regular 3-hydroxylation of the fatty acids during usual β-oxidation is expanded to 4- and 5- hydroxylations, the cytotoxic effect of free fatty acids in the nutrient medium is alleviated or at least reduced. As a result, high concentrations of triglyceride in the media may stimulate the formation of lactones. Further measures towards high yields are a suitable design of the bioreactor and peripheral devices, and again the *in situ* removal of product.

5 Genetic Approaches

The examples cited demonstrate that several laboratories are now proceeding from incidental success to more concerted strategies of the type that have been applied for some years in the study of bioprocesses for therapeutically valuable compounds. The inevitable steps are an extended screening of strains, systematic optimization of culture conditions, and continuous strain improvement. Looking back on the history of the manufacture of antibiotics, it took decades to develop processes and strains to the present high level. Based on the accumulated knowledge, processes for the generation of the bioflavours will be developed much faster, but clear research goals and strategies will be needed. Five out of the following seven examples for a concerted genetic approach originate from Japanese laboratories.

The degradation of terpenic alcohols such as geraniol that are highly abundant in many plant essential oils can be directed to yield products such as 6-methyl-5-hepten-2-one, a compound with a pleasant fruity smell. A US patent[46] showed that genetically modified strains of *Pseudomonas putida* catalyse this reaction. Specific plasmids were transferred into the active biocatalyst.

[43] J. Sarris and A. Latrasse, *Agric. Biol. Chem.*, 1985, **11**, 3227.
[44] R. Cardillo, C. Fuganti, G. Sacerdote, M. Barbeni, P. Cabella, and F. Squarcia, Eur. P. 356 291, 1990.
[45] G.F. Kapfer, R.G. Berger, and F. Drawert, *Chem. Mikrobiol. Technol. Lebensm.*, 1990, **13**, in press.
[46] Microlife-Technics, US. P. 224 717, 1988.

Most of the genetic work on aroma relevant strains refers to yeast. Mating[47] or fusing[48] appropriate strains with supplementing properties were reported to improve the flavour of wine and sake, respectively. Mutants resistant to fluorophenylalanine isomers were selected and shown to be overproducers of phenylethanol and its acetate.[49] When used for the fermentation of sake a beverage with characteristic flavour notes was obtained. The same objective was achieved by using another mutant capable of producing hexanoic acid and, consequently, ethyl hexanoate,[50] and by a further mutant that took up increased amounts of leucine due to a defect in arginine permease.[51] Double mutants that were blocked in ergosterol synthesis and defective in alcohol dehydrogenase were patented for the production of sparkling wine.[52] This mutant accumulated and secreted geraniol, farnesol, and linalool resulting in a distinct change of the flavour profile of the fermented beverage.

The latter example also demonstrates that a detailed insight into the underlying metabolic routes and their key enzymes will be the essential background for any reasonable approach to establish further high yielding cultures and bioprocesses in the future.

[47] S. Iino and M. Watanabe, *Ferm. Bioind. Chem.*, 1989, **3**, 69.
[48] K. Takahashi and K. Joshizawa, in 'Distilled Beverage Flavour', ed. J.R. Piggott and A. Paterson, Ellis Horwood, Chichester, 1989, 309.
[49] K. Fukuda, M. Watanabe, K. Asano, H. Ueda, and S. Ohta, *Agric. Biol. Chem.*, 1990, **54**, 269.
[50] E. Ichikawa, J. Hata, S. Imayasu, and K. Suginami, Jap. P. 63 309 175, 1988.
[51] O. Akita, T. Hasuo, S. Hara, and K. Joshizawa, *Hakko Kogaku Kaishi*, 1989, **67**, 7. [*Chem.Abst.* 1989, **110**, 133588.]
[52] F. Karst and B.D.V. Vladescu, Eur. P. 313 465, 1989.

Flavour Improvement in Apples and Pears Through Plant Breeding

F.H. Alston

HORTICULTURE RESEARCH INTERNATIONAL, EAST MALLING, WEST MALLING, KENT ME19 6BJ, UK

1 Introduction

Plant breeders rarely aim specifically to improve flavour; they usually plan to make overall improvements to a crop. Characters such as disease resistance and yield that can be assessed objectively in comparison with existing varieties, are frequently emphasized. Flavour, which is in most cases the result of a complex chemical relationship, is difficult to assess routinely in the large populations necessary for successful breeding programmes and it is therefore frequently considered at a late stage, when more easily assessed characters have been accounted for; even the best flavoured new selection will not be a success without high cropping efficiency and a good level of disease resistance. Breeding efficiency is determined by a number of factors, most of which are related to the genetic make-up of a crop and the development of efficient screening procedures. Characters that have a high heritability and for which a rapid screening technique has been developed, are easiest to manage. Frequently, effective progress is possible through selection for a particular character, without a full understanding of its genetics, physiology, or chemistry.

Most of the flavours appreciated today in plant products were recognized many years ago and are often best represented in old varieties unsuited to large scale commercial production. In addition, in fruits for example, these assessments were usually based on impressions of ripe products at harvest time. It is difficult at that stage to determine specific flavour characteristics which will be sufficiently stable after long storage and marketing periods.

In some genera the development of successful crop plants is a result of the selection of palatable variants by early man, *e.g. Solanum* where most species carry toxic substances such as glycoalkaloids.[1] Certain significant flavour components have been found to be determined by single genes such as the bitter principle in almonds (*Prunus amygdalus*), due to the presence of the glucoside amygdalin[2] and the 'burp' factor in cucumbers (*Cucumis sativus*) determined by the presence of curcubitacins.[3] Off-flavours in soybean (*Soja max*) products have been related to lipoxygenase oxidation of fatty acids, and recent work indicates that selection for absence of the lipoxygenase-2 isoenzyme may reduce many of these unwanted characters.[4] These examples show that, although it is often difficult to measure flavour objectively, it is possible to make significant advances in some crops through selection for the presence or absence of individual compounds.

Characteristic flavours in many crops depend on relative levels of sugars and acids. Improvements in tomato (*Lycopersicon esculentum*) flavour have largely been achieved through selection for increased sugar and acid levels.[5]

Volatile compounds, contributing to characteristic aromas, are important components of flavour. In some cases distinct flavour differences have been identified as due to particular compounds, such as oct-1-en-3-ol in the snap bean (*Phaseolus vulgaris*), determined by a single gene.[6] In other cases a large number of volatile compounds are involved and it is difficult to determine which are the most important. It is usual to find that there is a paucity of identifiable qualitative differences, which reflects the complex origin of such flavours.

Sometimes selection for flavour components is complicated through negative correlations with components of yield. An example of this, related to plant habit, is found in the selection for high yield through determinate growth in tomato which is said to lead to the production of fruit with low sugar content,[7] and a similar situation has also been reported in dwarf peaches (*Prunus persica*).[8] In both crops the special growth habits were chosen as producing more efficient cropping systems, increasing the proportion of fruiting tissues to vegetative tissue.

Fruit crops, the main concern of this chapter, may feature as raw materials in many processed products but since they are usually consumed as fresh products the eating quality is a critical factor in the marketing of new varieties. Invariably, in the case of these crops, the most effective selection is based on sugar and acid content, the most favoured being described as a moderate to low acidity combined with a medium to high sugar content.[9] Texture, which is regarded as a critical complement to flavour in many crops, can be measured instrumentally. It is, however, the interaction of

[1] T. Johns and J.G. Alonso, *Euphytica*, 1990, **50**, 203.
[2] M.J. Heppner, *Genetics*, 1923, **8**, 390.
[3] J.M. Andeweg and J.W. De Bruyn, *Euphytica*, 1959, **8**, 13–20.
[4] C.S. Davies, S.S. Neilsen, and N.C. Nielsen, *J. Am. Oil Chem. Soc.*, 1987, **64**, 1428.
[5] R.A. Jones and S.J. Scott, *HortScience*, 1984, **109**, 318.
[6] M.A. Stevens and W.A. Frazier, *Proc. Am. Soc. Hort. Sci.*, 1967, **91**, 274.
[7] M.A. Stevens, *Plant Breeding Rev.*, 1986, **4**, 273.
[8] P.E. Hansche, *HortScience*, 1986, **21**, 1193.
[9] C. Fideghelli, *Genet. Agr.*, 1988, **42**, 105.

taste, texture, and aromatic properties which determines the ultimate acceptability of new selections.

2   Improving Apple and Pear Flavour

2.1   Development and Perception

Although a combination of acids, sugars, and volatiles forms the basis of flavour in the apple (*Malus pumila*), variations in flesh texture have a significant effect on flavour appreciation. Flesh colour and skin colour also modify the consumer's anticipation of apple flavour. While it is feasible to define desirable acidity and sugar levels, in most cases precise assessments of flavour are only possible through taste panels. A wealth of special flavours has been identified amongst old apple varieties, outstanding examples being the aniseed flavour of 'Ellison's Orange', the 'nutty' flavour of 'Blenheim Orange' and the distinctive aromatic flavour of 'Cox's Orange Pippin'. Most of these varieties have serious commercial drawbacks, as their fully deserved reputations depend on assessments at optimum periods of development. Thus, 'Blenheim Orange' can rapidly become dry and crumbly after harvest, the perfumed and sweet flavoured variety 'American Mother' tastes like cotton wool after a cold summer, and the variety with the best reputation for aromatic flavour, 'D'Arcy Spice', has often to be left on the tree until early December in order to achieve the expected flavour.[10]

Little is known of the evolution of dessert apples and pears but the first stage was probably the selection of fruits without tannins, followed by the selection for palatable blends of sugar and acidity, leaving the more acid types for cooking. It seems that today's buttery sweet pears (*Pyrus communis*) developed similarly from tough astringent types. In this case a definite breeding and selection process is documented, carried out in Belgium and France during the eighteenth and early nineteenth century.[11] In Europe, in the late nineteenth century and the early twentieth century, apples were predominantly acidic. Later, North American varieties with low acidity were grown and introduced into European breeding programmes. Varieties such as Golden Delicious made important contributions to the development of more palatable apples for Europe, not only through reducing acidity levels but also by introducing crisp and juicy texture, which combined in crosses with European varieties with high soluble solids to produce some superb varieties for present day markets, which store until early spring. Thus an improvement in apple flavour over an extended season has resulted from breeding and selection, but in many respects the full potential has not yet been achieved.

The success of the new varieties depends on the response of the consumers. Current views in Northern Europe suggest that the public taste is turning

---

[10] J. Morgan, *The Garden*, 1982, **107**, 308.
[11] U.P. Hedrick, 'Report of the New York Agricultural Experiment Station for the year 1921, Vol. II', pp. 636.

from moderately acidic aromatic apples, such as 'Cox', to sweeter sub-acid varieties like 'Elstar' and 'Jonagold', both derivatives of 'Golden Delicious'. Surveys in England show a market divided into two parts, those preferring sweet apples and those preferring sharp or acidic apples. It also appears that it is important for apples to 'look' as though they have a good flavour. These observations are based on consumer perceptions of the current mass market apples. Red colour is associated with sweetness ('Red Delicious'), green skin with sharpness and crispness ('Granny Smith') and partly red striped apples on a green background with moderate acidity and a 'nice' flavour ('Cox'). Breeding experiments show that these skin colour features segregate independently of flavour components. There is little evidence that many consumers perceive the special aromatic component of Cox, though a few comment on its 'nice' or 'delicious' flavour.

2.2   Flavour Breeding Targets

2.2.1   *Acidity and sugar content*

Malic acid which forms the most important component of apple flavour is controlled by a single dominant gene (*Ma*).[12] Most dessert varieties are heterozygous for this gene, and hence most progenies show distinct segregations for acidity. The 25 % or so of seedlings in these progenies which have the homozygous genotype *ma ma* have an unacceptably low acidity which leads to a bland flavour. Most such low acid (0.0–0.4 %) seedlings are discarded, except for certain very early harvested selections where low acidity is an asset. Amongst cider varieties, 'bitter sweet' varieties with very low acidity and high phenolics, form an important group. A quantitatively determined inheritance pattern is superimposed on the *Ma* segregation pattern,[13] its most important effect being on the distribution of acidity levels amongst the acidic group (0.5–1.5 %); desirable dessert apples ranging from 0.5 % to 0.9 %. Thus in progenies where one of the parents is highly acidic, a large proportion of plants will be discarded through having too high acidity.[14] For example, in a cross between the very acidic dessert variety James Grieve (1.0 %. pH 3.1) and the sub-acid variety Golden Delicious (0.4 %, pH 3.5), only 11 % of seedlings fell into the acceptable acidic category (pH 3.4–3.6). It is most convenient and sensible to assess acidity on the basis of pH where large numbers of samples are involved, since the perception of acidity is considered to be dependent on the hydrogen ion concentration.

Fructose, sucrose, and glucose are the principal sugars in apples, with fructose often accounting for 50 % of the total sugar. The amount of these substances is inherited quantitatively with the progeny means approaching the mid-parent values.[13] Acidity and sweetness are inherited independently, but the balance of sugar content and acidity determines acceptability.

[12] N. Nybom. *Hereditas, Lund*, 1959, **45**, 332.
[13] A.G. Brown and D.M. Harvey, *Euphytica*, 1971, **20**, 68.
[14] T. Visser and J.J. Verhaegh, *Euphytica*, 1978, **27**, 753.

Highly acidic/low sugar apples are usually unpalatable, and successful dessert apples fit into the moderate acidity (0.5–0.9 %)/moderate sugar (12–13 %), moderate acidity/high sugar (14–16 %) and low acidity (0.0–0-.4 %)/moderate sugar categories.

In apples we seek a pH of 3.4 or 3.5 with a sugar content of 14–15 %, while in pears we aim for a lower acidity (pH 4.5). Pear progenies are mainly distributed within the range covered by the low acidity sector of apple progenies and they have a relatively low sugar content, 10–12 % according to season. In view of this low sugar content it is desirable to select for those in the lower portion of the acidity distribution, in order to achieve a satisfactory blending of the two components, on a background with a characteristic pear flavour.

2.2.2   Volatile components

Following empirical assessments for good aroma based on tasting derivatives of 'Cox', the source of good flavour, it is possible to classify seedlings into two distinct groups, those with an aroma similar to 'Cox' and those without. This distinctive aromatic component segregates as if it were determined by a single dominant gene.[15] The special flavour of the 'Cox'-type selections produced from the breeding programme at East Malling were selected this way. 'Cox' appears to be heterozygous for its distinctive aromatic component. The number of volatile components recorded in the apple exceeds 250.[16]

In an attempt to find which volatiles are significant aromatic components in 'Cox' flavour, two gas chromatographic peaks were found to show a particularly striking correlation in a 'Cox' selfed progeny.[17] These peaks were later related to hexylacetate and hexylbutyrate, which have been identified as important flavour components both in 'Golden Delicious' and 'Cox' and hence should probably be regarded as important flavour impact compounds since they are clearly not variety specific. It has not yet been possible to find a single volatile component related to 'Cox' flavour that can be usefully identified and screened for in apple progenies. However, the less important but characteristic aniseed component of the apple 'Ellison's Orange' has been identified as 4-methoxyallylbenzene[16] and merits genetic studies.

Breeding and selection for specific volatiles has been developed further in pears, using spectrometry. The outstanding flavour from 'William's Bon Chrétien' ('Bartlett') in canned fruit was found to be associated with a high level of decadienoate esters.[18] In a survey of fresh fruit, three varieties 'Colmar d'Eté', 'Catillac' and 'Merton Pride' were found to have 'William's'

[15] F.H. Alston and R. Watkins, 'Proceedings Eucarpia Fruit Section Symposium V', Top Fruit Breeding, Canterbury, 1973, p. 14.
[16] A.A. Williams, O.G. Tucknott and M.J. Lewis, J. Sci. Food Agric., 1977, 28, 185.
[17] D.F. Meigh, 'Report of the Ditton Laboratory for 1967–68', p. 53.
[18] H.A. Quamme and P.B. Marriage, Acta Hort., 1977, 69, 301.

type aroma and high decadienoate ester levels. These levels were either very low or absent in nine other varieties without 'William's' aroma.[19] This information has been applied in a pear breeding programme aimed at producing high quality processing varieties resistant to fire blight. The new variety 'Harvest Queen' produced from this programme shows the same decadienoate ester levels as 'William's' and the quality of the processed product is similar to that of 'William's'.[20]

The techniques applied in these studies of aromatics are not as suitable as rapid screening techniques for plant breeding as are the application of the pH meter and the refractometer for acid and sugar assessment. More rapid techniques need to be developed; further genetic studies might result in convenient marker genes being located that can be used as assays for flavour at an early stage. Variations in fruit maturity between selections also handicap accurate flavour comparisons, particularly the measurement of volatile components, the production of which is inhibited during storage and prior to maturity. Difficulties in identifying varietal flavour characteristics by gas chromatography may also be a result of some flavours being due to trace amounts of compounds for which the mouth is very sensitive, whereas the mouth may be less sensitive to some compounds such as hexyl and butyl acetate which could be present in greater quantities.

2.3   Breeding Procedure

In common with other vegetatively propagated crops, new apple and pear cultivars are usually derived from segregating F_1 generations. However, as tree crops they need to pass through a non-productive or juvenile period prior to cropping, which can vary from 4 to 15 years. The juvenile period delays the selection of new varieties and also the testing and establishment of commercial orchards of new varieties, since it is closely related to the length of the productive phase of adult trees. There is a relatively high additive variance governing the inheritance of most traits; hence progeny means can often be predicted from mid-parent values. However, several important commercial characters are determined by single dominant or recessive genes, which in addition justifies a careful matching of parents according to genotype. Individuals, chosen as exceptional members of progenies produced in this way, are selected and tested in orchard trials.[21]

Market and consumer requirements have special effects on apple and pear breeding policy in which fruit quality forms the prime target.[22] Cropping, harvesting periods, durability in storage, and shelf life have to be taken into account as well as flavour. Disease and pest resistance are recognized today as being of particular importance to consumers; their potential

[19] H.A. Quamme, 'Report of East Malling Research Station for 1980', p. 110.
[20] H.A. Quamme and G.A. Spearman, *HortScience*, 1983, **18**, 770.
[21] F.H. Alston and P. Spiegel-Roy, in 'Attributes of Trees as Crop Plants', ed. M.G.R. Cannell and J.E. Jackson. Institute of Terrestrial Ecology, Huntingdon, 1985, p. 49.
[22] F.H. Alston, in 'Quality in Stored and Processed Vegetables and Fruit', ed. P.W. Goodenough and R.K. Atkin. Academic Press, London, 1981, p. 93.

for reduction of agrochemical inputs is likely to be very significant.[23] It is also important to develop new varieties which will store without the need for post harvest dips to control fungal diseases and physiological disorders.[24]

The combination of good flavour with high yield, long storage, and resistance to pests and diseases is hampered by the necessity to conduct several crosses to combine all features. Priorities must be decided on the basis of feasibility. In a programme planned to combine seven principle characters, yield, skin finish, fruit colour, fruit size, acidity, storage performance, and texture to the standards required in a new commercial apple variety, the initial progeny size should be at least 10 000 in order to have a good chance of success.[25]

2.4   Improving Flavour After Storage

There is a great potential for extending the marketing season of high quality apples through breeding and selection. Naturally late maturing varieties are frequently not suited to marketing, as they frequently lack a crisp texture, a suitable sugar/acid balance, and an aromatic flavour after long term storage under modern conditions. Acidity can be reduced during storage although controlled atmosphere storage in low oxygen concentrations can arrest this decline; it can also result in a depletion of important flavour compounds such as butyl and hexyl acetate.[26] Careful selection for flavour after long term storage has resulted in new varieties such as 'Malling Fiesta' ('Cox' × 'Idared'). This retains much of its 'Cox type' aroma after storage until mid June in combination with the long storage properties and crisp texture of 'Idared', a variety with a poor flavour, which apparently lacks desirable volatile components.

The stringent storage selection process starts on the first fruits of seedlings in the breeding plots. It is conducted by using a series of sealed plastic bins, ventilated with moist air to maintain humidity levels comparable to commercial conditions.[24] At that stage a simple storage regime is chosen, 0 °C in air. Although many selections are susceptible to low temperature breakdown in such conditions, the aim is to select new varieties which will store well in conditions which are less sophisticated than current controlled atmosphere regimes. Flavour assessment is made when apples are removed from store. A relatively low proportion of seedlings survive these early tests, emphasizing the necessity for large progenies to allow also for sufficient improvements in characters other than storage performance and flavour.

[23] F.H. Alston, V.H. Knight, and D.W. Simpson, in 'Control of Plant Diseases: Costs and Benefits', ed. B.C. Clifford and E. Lester, Blackwell, Oxford, 1988, p. 67.
[24] F.H. Alston, *Acta Hort.*, 1988, **224**, 177.
[25] F.H. Alston, in 'Improving Vegetatively Propagated Crops', ed. A.J. Abott and R.K. Atkin, Academic Press, London, 1987, p. 113.
[26] M. Knee and R.O. Sharples, in 'Quality in Stored and Processed Vegetables and Fruit', ed. P.W. Goodenough and R.K. Atkins, Academic Press, London, 1981, p. 93.

2.5 Pre-selection for Flavour

Rapid and meaningful screening of large progenies at the earliest possible stage is a requirement in all plant breeding programmes. It is particularly important in tree groups which can occupy a large area and have long infertile periods prior to cropping.

Early selection techniques have been developed for yield, disease resistance, and tree type, which can be applied during the two years following germination of the seed.[25] Some 26 agronomic genes have been identified in apple and over 30 isoenzymic genes[27] which have the potential to act as marker genes, useful in early selection. At the moment resistance genes[27], a growth gene[28], and a self incompatibility locus[29] have been tagged in this way. Studies are underway on quality components including flavour. Early selection processes for flavour depend on achieving rapid fruiting of selections after budding onto dwarfing rootstocks, or developing growing systems aimed at rapid growth, both processes being designed to shorten the juvenile phase.[30]

Rapid classification of the first fruits into acid and sweet groups can be made before harvest using the standard indicator bromo-cresol green on the cut surface of fruits, or a more detailed result can be achieved by using indicator papers. When fruit is assessed after storage the apple flesh is penetrated using a bayonet probe electrode in conjunction with a pH meter.

There are good prospects of selecting for fruit acidity prior to fruiting in the year after seed germination. Leaf sap pH is correlated with fruit pH[31] and, provided information is available on parental phenotypes, is potentially useful for mass screening of large populations. The search continues for more useful pre-selection criteria at the young seedling stage.

Work has also started on investigating restriction fragment length polymorphisms in apple, flavour genes and quantitative linked traits being important targets.

3 Conclusion

Most plant breeding programmes expect to maintain flavour standards rather than improve them, using as reference bases old or popular varieties.

However, modern marketing and storage developments create a need for new varieties to be able to survive bulk storage in modified atmospheres, followed by rapid grading and packaging for display in cooled cabinets, and still be able to carry a good flavour to the consumer. It is important for plant breeders to develop flavour studies, to identify flavour constituents and to

---

[27] A.G. Manganaris, 'Isoenzymes as Genetic Markers in Apple Breeding', PhD Thesis, University of London, 1989.

[28] A.G. Manganaris and F.H. Alston, *Acta Hort.*, 1988, **224**, 177.

[29] A.G. Manganaris and F.H. Alston, *Theor. Appl. Genet.*, 1987, **74**, 154.

[30] F.H. Alston and K.R. Tobutt, in 'Manipulation of Fruiting', Butterworth, London, 1989, p. 329.

[31] T. Vissar and J.J. Verhaegh, *Euphytica*, 1978, **27**, 761.

determine their genetic basis. Early location of such genes through genetic markers is necessary in order to advance breeding programmes and to create the potential, by the transference and recombination of key genes, to improve or create new flavours. Such careful combination of flavour constituents by breeding, selection, and genetic manipulation opens the potential for the adaptation of metabolic systems related to flavour to various environmental conditions.

There is a potential for producing new apple varieties for the Northern Hemisphere which will store well, without post-harvest dips, for marketing between May and August. Such varieties would be derived from recent varieties which combine good texture, skin finish, and fruit size with good flavour; very few of the old, naturally late maturing varieties are expected to be useful in this respect. Currently, breeding programmes emphasize pest and disease resistance, following previous success in breeding for increased yield. Future emphasis will be on flavour and other quality attributes, including the introduction of the full range of desirable flavours available amongst apple and pear varieties into new commercial varieties with high yield, disease and pest resistance, and adaptation to particular environmental conditions.

Acknowledgements

I am indebted to my colleagues Dr R.O. Sharples and Mr K.R. Tobutt for their helpful comments and criticism during the preparation of this chapter.

Flavour Improvement Through Plant Cell Culture

M.J.C. Rhodes, A. Spencer, J.D. Hamill*, and R.J. Robins

PLANT BIOTECHNOLOGY GROUP, GENETICS AND MICROBIOL-
OGY DEPARTMENT, AFRC INSTITUTE OF FOOD RESEARCH,
COLNEY LANE, NORWICH, NR4 7AU, UK

1 Introduction

An important requirement for the rational improvement of flavour compound production in plants is an understanding of the pathway and enzymology of the biosynthesis of particular products. Such knowledge offers
the opportunity to manipulate pathways by genetic means to increase the
formation of particular desirable components and to reduce or eliminate
undesirable ones. Our knowledge of plant secondary metabolic pathways,
including those leading to flavour compounds, is frequently very limited.
These pathways are often difficult to study in depth in whole plants where
the biosynthetic sequence may only be expressed in particular cell types
within a specific plant organ and even then only at a certain time of the year.
Where progress is being made, the availability of tissue culture systems with
stable growth and production of the target compounds is a major factor.
Such systems allow experimentation throughout the year, give high rates of
product synthesis, provide axenic systems for radiolabelling and inhibition
studies, and are good sources for the extraction of enzymes for *in vitro*
studies. The availability of stable, productive tissue culture systems is a
major element in recent studies in which the complete pathway for
berberine biosynthesis in *Berberis*[1] and in *Coptis*[2] has been described.

* Present address: Department of Genetics and Developmental Biology, Monash University,
  Clayton, Melbourne, Australia
[1] M.H. Zenk, in 'The Chemistry and Biology of Isoquinoline Alkaloids', ed. J.D. Phillipson,
  M.F. Roberts, and M.H. Zenk, Springer-Verlag, Berlin, 1985, p. 240.
[2] Y. Yamada and T. Hashimoto, in 'Applications of Plant Cell and Tissue Cultures', Ciba
  Foundation Symposium, No. 137, 1988, p. 199.

Plant-derived food flavours may be divided into a number of groups. Firstly there are the flavours based on endogenous components. *Simple flavours* are based on a single or small group of chemically closely-related flavour compounds: examples of simple flavours would include capsaicin in *Capsicum* and gingerols in *Zingiber*. *Complex flavours*, such as fruit flavours and essential oils, are those in which a considerable number of different compounds, often from different chemical classes, contribute to the flavour; examples would be mint and citrus oils, and strawberry and raspberry fruit flavours.

Secondly, there are the flavours which develop from endogenous non-flavour precursors during post-harvest treatment. This may be during eating when, for example, flavours such as cucumber, garlic, onion, and Brassica[3] arise by enzyme action on the flavour precursors. In the onion, *S-trans*-prop-l-enyl-L-cysteine sulphoxide together with the related *S*-methyl and *S*-propyl derivatives are the main non-volatile flavour precursors: during cell breakage the precursors are exposed to an enzyme, alliinase, which causes the sulphoxide precursors to yield sulphenic acids together with ammonia and pyruvate. The sulphenic acids subsequently undergo a series of chemical reactions to yield a range of volatile sulphur-containing flavour compounds. A further type of secondarily derived flavours are those that develop from endogenous non-flavour components to yield the typical flavour during a fermentation or other process. These flavours include the development of cocoa and vanilla flavour during fermentation, and of coffee flavour during roasting.

This chapter will briefly review examples where progress in the development of tissue culture systems for the formation of flavours has been made and will consider the problems encountered with more complex flavours. Some of our own recent work to develop transformed organ cultures which may allow more complex flavours to be tackled by tissue culture methods will also be discussed.

2 Use of Plant Cell Cultures for Flavour Compound Formation

A number of different types of plant cell culture, callus, cell suspensions, and immobilized cell cultures have been studied for the production of flavour compounds.[4, 5, 6] An example of a simple flavour which has been studied in culture is capsaicin, the pungent principle of chilli pepper. Cell suspension cultures of *Capsicum frutescens* which produced low levels of

---

[3] P. Schreier, in 'Development In Food Flavours' ed. G.G. Birch and M.G. Lindley, Elsevier, London, 1986, p. 89.
[4] Y.C. Hong and S.K. Harlander, in 'Flavour Chemistry of Lipid Foods', ed. D.B. Min and T.H. Smouse, AOCS, Champaign, Illinois, 1989, p. 348.
[5] D. Knorr, M.D. Beaumont, C.S. Caster, H. Dorenburg, B. Gross, Y. Pandyai, and L.G. Romagnoli. *Food Technol.*, 1990, **44**, 71.
[6] H.A. Collin and M. Walls, in 'Handbook of Plant Cell Culture' Vol. 1, ed. D.A. Evans, W.R. Sharp, P.V. Ammirato, and Y. Yamada, Macmillan, 1983, p. 729.

capsaicin were developed by careful selection of cell lines and of medium conditions.[7] The productivity of these cultures was increased by more than 100-fold by immobilization of the cells in reticulated polyurethane foam particles.[8] Further increases in productivity were obtained by feeding isocapric acid, one of the immediate precursors of capsaicin, to cultures, and a maximum specific production rate of 0.5 mg capsaisin g dry weight^{-1} d^{-1} has been reported.[9] Scale-up of these cultures has been undertaken using beds of stationary foam particles in which the *Capsicum* cells are immobilized.[9]

A chemically-related food flavour, namely vanilla, is formed from endogenous glycosidic precursors by action of a fruit β-glucosidase during a fermentation process. Although vanillin is the major flavour component of vanilla flavour, fermentation yields other minor flavour compounds which impart a superior flavour to natural vanilla beyond the major note given by vanillin. There is great interest in developing tissue culture systems to produce natural vanilla flavour compounds. Knuth and Sahai[10] have developed cell suspension cultures of *Vanilla planifolia* which produce up to 383 μg l^{-1} d^{-1}, although the product compound exists as the free phenol in culture rather than as the glycoside. Romagnoli and Knorr[11] developed callus cultures of *Vanilla planifolia* whose production of vanillin was stimulated by feeding a potential precursor, ferulic acid. The role of ferulic acid as a precursor of vanillin has been challenged by Funk and Brodelius[12-14] in a series of recent papers. Their culture of *Vanilla* had no accumulation of phenolic flavour compounds unless it was elicited with chitosan which induced the accumulation of vanillic acid rather than vanillin. They have used precursor feeding and inhibition experiments, together with description of the specificities of key enzymes to suggest that isoferulic acid rather than ferulic acid is the precursor of vanillic acid (and presumably of vanillin) in their culture.

Flavour formation in onion and garlic involves the action of alliinase on *S*-alkyl-cysteine sulphoxide derivatives and thus both components must be present in cell culture for successful flavour development. Cultures and cell suspension cultures of onion (*Allium cepa*) have been developed by a number of groups.[4] Selby *et al.*[15] showed that their cultures failed to produce onion flavours but could be stimulated to do so if fed with the immediate flavour precursors suggesting that the enzyme alliinase was present in the cultures. The presence of the enzyme but absence of the precursors appears

[7] M.M. Yeoman, M.A. Holden, P. Corchet, P.R. Holden, J.G. Goy, and M.C. Hobbs, *Proc. Phytochem. Soc. Eur.*, 1990, **30**, 139.
[8] K. Lindsey and M.M. Yeoman. *J. Expt. Bot.*, 1984, **35**, 1684.
[9] F. Mavituna, A.K. Wilkinson, and P.D. Williams, in 'Bioreactors and Biotransformations' ed. G.W. Moody and P.B. Baker, Elsevier, London, 1988, p. 26–37.
[10] M.E. Knuth and O.P. Sahai. PCT Patent 1989 W089/00820.
[11] L.G. Romagnoli and D. Knorr, *Food Biotechnol.*, 1988, **2**, 93.
[12] C. Funk and P.E. Brodelius. *Phytochem.*, 1990, **3**, 845.
[13] C. Funk and P.E. Brodelius. *Plant Physiol.*, 1990, **94**, 95.
[14] C. Funk and P.E. Brodelius. *Plant Physiol.*, 1990, **94**, 102.
[15] C. Selby, I.J. Galpin, and H.A. Collin. *New Phytol.*, 1979, **83**, 351.

a common feature of many dispersed onion cultures. Alliinase has, indeed, been purified from onion cell cultures.[16] Experiments involving the feeding of ^{14}C-labelled *S*-(2-carboxypropyl)-L-cysteine, a precursor of the flavour precursor *S-trans* prop-l-enyl-L-cysteine sulphoxide, suggested that the final steps in the pathway are expressed even in cultures which are incapable of onion flavour generation.[17] This further suggests that the steps diverting primary metabolites such as valine and cysteine into formation of the flavour precursors, rather than limitations in the later steps in the pathway or in the availability of alliinase, limit flavour formation in these cultures. Musker *et al*.[18] have shown that low levels (0.5 mg l^{-1}) of the growth regulator picloram in the growth medium of onion callus stimulated accumulation of *S-trans*-prop-l-enyl-l-cysteine sulphoxide to levels of up to 40 % of that found in mature onion bulbs. These cultures have the capability of producing onion flavour by forming both of the essential components. Low levels of picloram induced root and leaf bud development, however, higher levels of picloram induced formation of the sulphoxide precursor without inducing differentiation and thus the two processes are not necessarily linked.[19] Recently other workers have been successful in inducing formation of cysteine sulphoxide derivatives in onion cultures.[20]

With more complex flavours, such as mint essential oil, cell cultures have been studied for a considerable period with only limited success. In early work[21] it was shown that callus cultures of *M. piperita* produced low levels of mint oil rich in pulegone and menthofuran compared with the parent plant where the total level of oil was much higher, and menthol and menthone were the dominant components. In a more recent study[22] of cell lines of *Mentha piperita* it was shown that monoterpenes represented only a minor component of the total volatile fraction. A number of different lines were established which were variable in appearance and structure. They, however, show only very low levels of monoterpene accumulation with trace amounts of menthone, neomenthol, and neomenthyl acetate and none of the lines accumulated menthol. The failure of mint cell cultures to accumulate monoterpenes has been attributed to rapid degradation of the monoterpenes rather than a failure to synthesize them.[24] The mechanism of breakdown has been studied in detail[23] (for a review see Reference 24). A further factor is that the monoterpene compounds are phytotoxic and in the

[16] S. Wright, Ph.D Thesis University of Sheffield, 1987.
[17] A. Turnbull, I.J. Galpin, and H.A. Collin. *New Phytol.*, 1980, **85** 483.
[18] D. Musker, H.A. Collin, G. Britton, and G. Ollerhead, in 'Manipulating Secondary Metabolism in Cultures' ed. R.J. Robins and M.J.C. Rhodes, Cambridge University Press, Cambridge, 1988, p. 177.
[19] H.A. Collin, D. Musker, and G. Britton, in 'Primary and Secondary Metabolism of Plant Cell Cultures II' ed. W.G.W. Kurz, Springer-Verlag, Berlin, 1989, p. 125.
[20] V.P. Dixit, Abs VII IAPTC Congress, Amsterdam, 1990, p. 328.
[21] J. Bricout and C. Paupardin, *Compt. Rend. Acad. Sci. Ser. D*, 1974, **278**, 719.
[22] F. Cornier and C.B. Do, in 'Bioflavour '87', ed. P. Schreier, W. de Gruyter, Berlin, 1988, p. 357.
[23] R. Croteau, in 'Flavours and Fragrances: A World Perspective', Proc. 10th International Congress of Essential Oils, Washington, Elsevier, Amsterdam, 1988, p. 65.
[24] J. Gerhenszon and R. Croteau. *Recent Adv. Phytochem.*, 1990, **24**, 99.

plant they are sequestered extra-cellularly in the space between cell wall and cuticle of the secretory cells of the oil glands.[24] In cell culture, there is no physical separation between synthesis and storage and no means of protecting nascent monoterpenes from immediate degradation. The use of two phase systems[25,26] has proven useful in promoting accumulation of terpenes in cell cultures of a number of species. Here lipophilic phases such as mixed triglycerides (myglyol)[25] or reverse phase GLC supports such as RP-8[26] have proved useful in adsorbing and stabilizing monoterpenes excreted into the medium. In mint, it was concluded that the presence of non-polar adsorbents was essential for detectable accumulation of monoterpenes.[22] In spite of the limited success with undifferentiated mint cultures, *M. piperita* callus which had been induced to differentiate adventitious shoots produced significant amounts of the normal mint monoterpenes.[27,28] This need for differentiation for mint oil accumulation will be considered further in Section 3.

Other complex flavours in which there is great interest are the fruit flavours, principally strawberry and citrus. Strawberry flavour is widely used in the food processing industry and development of a tissue culture system for strawberry flavour has been attempted by a number of workers[4,10] but progress has been limited. Strawberry flavour is complex containing over 200 components[29] including esters, ketones, alcohols, and organic acids and minor contents of lactones and furans. In one study[30] strawberry callus and liquid suspension cultures were established and it was shown[4] that only a limited number of flavour compounds were produced by liquid suspension cultures compared with the range of compounds present in the ripe fruit. A recent study on flavour compound formation in callus cultures of guava fruits[31] demonstrated that callus exhibited a perceptible guava fruit flavour under conditions of slow growth. The callus showed only 19 volatile compounds by GC analysis compared with 58 in the fruit tissue, and some major fruit flavour compounds were absent from the callus cultures.

From this brief review it can be seen that progress in developing cell culture systems for production of food flavours has been uneven. With simple flavours such as chilli pepper considerable progress has been made, while for the more complex flavours such as essential oils progress has been restricted. When many of the flavour components are monoterpenes, problems of the biochemical instability and phytotoxicity of the flavour compounds of interest are obstacles to progress. These problems are resolved in

[25] B.V. Charlwood, J.T. Brown, C. Moustou, G.S. Morris, and K.A. Charlwood, in 'Bioflavour '87', ed. P. Schreier, W. de Gruyter, Berlin, 1988, p. 304.

[26] H. Becker, J. Reichling, W. Bissan, and S. Herold, *Proc. Third Eur. Cong. Biotechnol.*, 1984, **1**, 209.

[27] J. Bricout and C. Paupardin, *Compt. Rend. Acad. Sci. Ser. D*, 1975, **281**, 383.

[28] J. Bricout, M.J. Garcia-Rodriquez, C. Paupardin, and R. Saussay. *Compt. Rend. Acad. Sci. Ser. D*, 1978, **287**, 611.

[29] S. Van Straten, 'Volatile Compounds in Food', 4th Edn., Central Institute for Nutrition and Food Research, Zeist, Netherlands, 1977.

[30] Y.C. Hong, P.E. Read, S.K. Harlander, and T.P. Labuzza, *J. Food Sci.* 1989, **54**, 388.

[31] T.N. Prabha, M.S. Narayanan, and M.V. Patwardhan, *J. Sci. Food Agric.*, 1990, **50**, 105.

the intact plant by compartmentalization and cell specialization. In the remainder of this chapter we will consider the prospects for organized cultures where many of these inherent problems of stability and productivity can be resolved. In particular the progress to develop transformed organ cultures will be discussed.

3 Use of Transformed Plant Organ Cultures for Flavour Compound Formation

Transformed organ cultures, which may be formed following infection of plant material with *Agrobacterium* vectors, offer the possibility of genetically and biochemically stable cultures without the need for exogenous supplies of hormones to sustain the culture. This is valuable since it is often observed with untransformed organ cultures, such as untransformed root cultures of *Hyoscyamus*,[32] that the levels of exogenous auxin required to optimize growth are antagonist to product formation. Further, it has been found that transformed root cultures have many advantages over comparable untransformed cultures such as higher growth rates, robustness, and enhanced capacity to grow in fermentation.[33] Additionally, transformed cultures offer direct approaches to the genetic transformation of the organ culture to influence growth and product formation. So far, however, the application of transformed organ cultures to flavour compound production has been limited. Initially the methodology was limited to the root-based flavours such as Angelica and liquorice but more recently the technology to develop transformed shoot cultures has been developed[34] and thus shoot derived-flavours may now be studied using this approach.

3.1 Transformed Root Cultures for Flavour Compound Production

There is increasing application of transformed root cultures for the production of valuable plant secondary products. The properties of such systems and their advantages over dispersed cell cultures have recently been reviewed[35, 36] These root cultures are formed following infection of plant material with the soil bacterium, *Agrobacterium rhizogenes* in a process which involves transfer of DNA from the bacterial plasmid into the plant genome. The transfer process is largely controlled by genes on the plasmid, the *vir* (virulence) genes which are not themselves transferred into the plant genome. The *vir* genes, produce products which excise part of the plasmid

[32] T. Hashimoto, Y. Yukimure, and Y. Yamada, *J. Plant Physiol.*, 1990, **124**, 61.
[33] M.J.C. Rhodes, R.J. Robins, J.D. Hamill, A.J. Parr, M.G. Hilton and N.J. Walton, *Proc. Phytochem. Soc. Eur.*, 1990, **30**, 201.
[34] A.J. Spencer, J.D. Hamill, and M.J.C. Rhodes, *Plant Cell Reps.*, 1990, **8**, 601.
[35] J.D. Hamill, A.J. Parr, M.J.C. Rhodes, R.J. Robins, and, N.J. Walton, *Bio/Technol.*, 1987, **5**, 800.
[36] M.W. Signs and H.E. Flores, *Bio Essays*, 1990, **12**, 7.

DNA (the so called t-DNA) delineated by 25 bp border sequences and effect its transfer and integration into the plant DNA. The t-DNA bears genes which are expressed in the eukaryotic environment and control both the factors (the *rol* genes) which induce the plant to form roots at the point of infection and those (*mps, ags*) which induce the plant to make specific bacterial substrates, the opines. The roots produced can be cleaned of free-living bacteria and maintained in *in vitro* axenic culture on simple media lacking plant hormones.[37] Such roots are frequently fast growing and produce the range of products characteristic of the plant species from which they were formed. The applications of such systems to flavour production have been limited. We have developed transformed root cultures of Angelica which we have shown to be transformed using polymerase chain reaction techniques and which produce many of the characteristic oil compounds (Robins and Rhodes unpublished data). Transformed root cultures of liquorice[38, 39] have been developed by two groups. In one study,[38] glycyrrhizin production was detected.

A long term interest of our group has been the study of the formation of the bittering agent quinine and related quinoline alkaloids in cultures of *Cinchona ledgeriana*. This product is extracted commercially from the bark of the tree, but evidence from various sources suggests that the capacity to synthesize these alkaloids is distributed widely in different organs of the plant.[40-42] In our experiments we have used a clone of *C. ledgeriana* (QC) supplied to us by Dr C.S. Hunter of Bristol Polytechnic and maintained as a sterile shoot culture. Table 1 shows that this material produces a range of methoxylated and unmethoxylated alkaloids but has only a low content of quinine. Cell suspension cultures were derived from this material as explant and in a growth promoting (2,4D/BA) medium yielded 188 µg l^{-1} d^{-1} of total alkaloid of which only 1.5 % was quinine.[43] This same culture in a production medium containing IAA and ZR yielded 984 µg l^{-1} d^{-1} of total alkaloids but only 13.8 µg of quinine l^{-1} d^{-1}.[43] Two transformed cultures were also developed from this same explant material. One, a crown gall culture,[44] rather similar in appearance to the cell suspension cultures, was developed following infection of the QC clone with *A. tumefaciens* strain A6. This yielded a culture in hormone-free medium which had similar total productivity (972 µg l^{-1} d^{-1}) to that of the cell suspension culture in the production medium: however, the yield of quinine (4.1 % of the total) was 2.8-fold higher. The second transformed culture was a root culture derived from

[37] J.D. Hamill, A.J. Parr, R.J. Robins and M.J.C. Rhodes, *Plant Cell Reps.*, 1986, **5**, 111.
[38] K.S. Ko, H. Noguchi, Y. Ebizuka and U. Sankawa, *Chem. Pharm. Bull.*, 1939, **37**, 245.
[39] K. Saito, H. Kaneko, M. Yamazaki, M. Yoshida, and I. Murakoshi, *Plant Cell Reps.*, 1990, **8**, 718.
[40] L.A. Anderson, A.T. Keen, and J.D. Phillipson, *Planta Med.*, 1982, **46**, 25.
[41] R.J. Robins, J. Payne, and M.J.C. Rhodes, *Planta Med.*, 1986, **52**, 220.
[42] J.D. Hamill, R.J. Robins, and M.J.C. Rhodes, *Planta Med.*, 1989, **55**, 354.
[43] M.J.C. Rhodes, J. Payne, and R.J. Robins, *Planta Med.*, 1986, **52**, 226.
[44] J. Payne, M.J.C. Rhodes, and R.J. Robins, *Planta Med.*, 1987, **53**, 367.

Table 1   Production of quinine and related quinoline alkaloids in a clone of Cinchona ledgeriana and in various types of plant tissue cultures derived from it

	Alkaloid production (μg g⁻¹ fwt)						Productivity (μg l⁻¹ d⁻¹)		Reference
	Cn*	Cd	Qd	Qn	Total	% Qn	Total Alkaloid	Quinine	
Intact Parent Plant (Clone QC)	53	2	0.4	5	60.4	8.2	–	–	41
1. Cell suspension Culture A. 2,4D/BA Medium[43]	12.2	5.6	1	0.3	19.2	1.5	188	2.8	43
B. IAA/ZR Medium[43]	80	5.3	7	2	14.2	1.4	984	13.8	43
2. Crown gall culture (A.tumefaciens A6)	65	2.0	1.2	3.7	90	4.1	972	38.9	44
3. Transformed root culture (A.rhizogenes LBA9402)	1.6	13.2	10.4	15.7	40.7	3.8	87	33	42

* Cn – Cinchonine; Cd – Cinchonidine; Qd – Quinidine; Qn – Quinine; 2,4D = 2,4-dichlorophenoxyacetic acid; BA = benzyladenine; IAA = indolylacetic acid; ZR = zeatin riboside.

infection of clone QC with *A. rhizogenes* LBA9402. This culture[42] was abnormal in structure compared with the typical transformed root structure and grew relatively slowly in culture. However it produced over 38 % of its total quinoline alkaloid as quinine and its overall quinine production 33 µg l^{-1} d^{-1} was only 15 % lower than that of the crown gall culture even though the total alkaloid production was less than 10 % of that of the crown gall culture. If lines of transformed roots of *C. ledgeriana* with growth rates comparable to typical hairy root lines[34] could be developed such systems might prove useful for the study of quinine formation.

Although transformed root cultures may prove useful in production of root-derived pharmaceuticals such as atropine, their application for food flavours is probably quite limited and this led us to investigate transformed cultures of other plant organs such as leaves and shoots where the opportunities for flavour compound production may be greater.

3.2   Transformed Shoot Cultures for Flavour Compound Formation

In contrast to *A. rhizogenes* which has evolved the capacity to induce root formation on a wide range of dicotyledonous plant species, there is no comparable species of *Agrobacterium* which normally has the ability to induce shoot formation after infection. *A. tumefaciens* induces crown gall formation following transformation. This bacterium carries a section of DNA (the t-DNA) on its Ti plasmid which is transferred into the plant genome using the products of the virulence genes in a manner essentially similar to that described for *A. rhizogenes*.[45-47] In *A. tumefaciens* strains producing the opine nopaline (such as strains T37, C58, N273) only one section of t-DNA is present in the Ti plasmid, but in strains producing octopine (*e.g.* A6NC, ACH5) two sections of t-DNA exist (t$_R$ and t$_L$) with t$_L$ having long regions of homology with the single t-DNA region of the nopaline strains.[46] Figure 1 shows a map of the t-DNA region of a nopaline strain indicating its delineation by 25 bp border sequences and the presence of various genes such as the gene controlling opine (nopaline) synthesis *ocs*, and three genes *tms1*, *tms2*, and *ipt* which appear primarily responsible for inducing the phenotypic response in the infected plant. These genes code for enzymes of auxin metabolism *i.e. tms1* for tryptophan mono-oxygenase, *tms2* for indoleacetamide hydrolase or of cytokinin biosynthesis *i.e. ipt* for dimethylallyl transferase. The presence and expression of these genes in the transformed plant tissue leads to over-production of auxins and cytokinins in the tissue.[48] Although other genes may play a modifying role, such as gene *6b*,[49] this over-production appears to lead

[45] K. Weising, J. Schell, and G. Khan, *Ann Rev. Genet.*, 1988, **22**, 421.
[46] L. Willmitzer, P. Dhaese, P.H. Schreier, W. Schmalenbach, M. van Montagu, and J. Schell, *Cell*, 1983, **32**, 1045.
[47] P. Zambyski, J. Tempe and J. Schell, *Cell*, 1989, **56**, 193.
[48] R.O. Morris, *Ann. Rev. Plant Physiol.*, 1986, **37**, 509.
[49] B. Tinland, B. Huss, F. Paulus, G. Bonnard, and L. Otten, *Mol. Gen. Genet.*, 1989, **219**, 217.

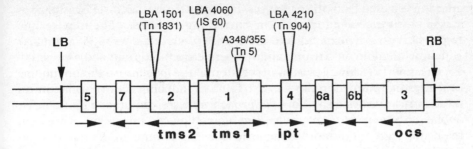

Figure 1   *A map of the* t-DNA *of a nopaline strain* (T37) *of* A. tumefaciens[46]

to enhanced cell proliferation and to gall formation. In tobacco, inactivation of either the auxin or cytokinin genes by transposon insertion leads to either shoot or root rather than gall formation.[50, 51] Inactivation of the auxin genes leads to shoot formation, while inactivation of the cytokinin gene leads to root formation. These effects can be reversed to give gall formation if the resultant shooty tissue is treated with auxins or the rooty tissue treated with cytokinins.[51] The so called 'shooty' mutants of *A. tumefaciens* with inactivated auxin genes provide an approach to developing transformed shoot cultures in other species. Further, if inactivation of the *tms* genes is all that is required for transformation then an additional approach is to express the *ipt* under its own promoter in the absence of the auxin genes. The construct pSSI[52] has the *ipt* gene of *A. tumefaciens* T37 under the control of its own promoter in pBIN19 housed in a disarmed strain of *A. tumefaciens* LBA 4404 from which the t-DNA has been excised. In the case of tobacco transformed with either the 'shooty' or the 'rooty' mutants, there is evidence that inactivation of either the auxin or the cytokinin genes changes the balance of auxins and cytokinins in the tissue, and that high endogenous cytokinin to auxin ratios are associated with the development of shooty cultures in tobacco.[53]

A further approach to induction of shoot cultures in a wider range of species was suggested by observations[47,54] that certain nopaline strains of *A. tumefaciens* induced phenotypes other than the expected galls following infection of certain tissues. Gresshoff *et al*[55] observed a low incidence of formation of shooty teratomas on *N. tabacum* cv Havana transformed with the nopaline strain, T37, C58, and N273 on shooty teratoma formation in a range of species.

A final approach which we took was to make constructs in which the *ipt*

50 D.J. Garfinkel, R.B. Simpson, L.W. Ream, F.F. White, M.P. Gordan, and E.W. Nester, *Cell*, 1981, **27**, 143.
51 G. Ooms, P.J.J. Hooykaas, G. Moolenaar, and R.A. Schilperoot, *Gene*, 1981, **14**, 33.
52 S. Schofield, personal communication.
53 D.E. Akiyoshi, R.O. Morris, R. Hinz, B.S. Mischke, T. Kosuge, D.J. Garfinkel, M.P. Gordon, and E.W. Nester, *Proc. Nat. Acad. Sci., USA*. 1983, **80**, 407.
54 G. Ooms, P.J.J. Hooykaas, G. Moolenaar, and R.A. Schilperoot, *Gene*, 1981, **14**, 33.
55 P.M. Gresshoff, M.L. Skotnicki, and B.G. Rolfe, *J. Bacteriol.*, 1979, **137**, 1020.

gene was isolated from the t-DNA of T37 and placed under the control of a range of available promoters of increasing strength. These were the nopaline synthase promoter and the cauliflower mosaic virus 355 promoter with or without the upstream enhancer sequence.[56] This enabled us to study the effect of increasing levels of expression of the *ipt* gene on the phenotype of the resulting transformants in the hope of inducing shoot formation on recalcitrant species. The four approaches described above have been applied to a range of plant species producing flavour compounds. However, the main target of this work is the genus *Mentha* and the production of mint oils. Four species of mint were tested, but two [*M. piperita citrata* ('*M. citrata*' Bergamot mint) and *M. piperita vulgaris* ('*M. piperita*' Scotch Black Mint)] which produce very distinct spectra of monoterpenoid oil components were chosen for more detailed study.

3.2.1 Use of 'shooty' mutants of A. tumefaciens

Tables 2 and 3 show the phenotypes induced by the *Agrobacterium* mutants on two model species, *Nicotiana tabacum* and *N. rustica*, and on seven flavour-producing plant species. As show in Table 2, the wild type strain, A348, induced the predicted gall formation on all the species tested. Six 'shooty' mutants having insertion of the transposon Tn5 (see Figure 1) in the *tms1* gene showed the expected phenotype on *N. tabacum* with either direct form-ation of shoots from the wound site or with shoots developing immediately after initial callus formation. However, even in the closely related species *N. rustica* all but one of the mutants failed to induce shoot formation. None of the herb species reacted by forming shoots: most produced galls, some did not react at all to some of the mutants and in one case (fennel) with *tms* 337::Tn5 a gall showing root development resulted. Table 3 shows similar data for another series of mutants, those based on the octopine strain Ach5. In this case mutants with insertions in both *tms1* and *tms2* were available and they both gave the predicted shooty cultures in *N. tabacum*, but again neither *N. rustica* nor any of the seven herb species tested gave the expected phenotype. In contrast, the 'rooty' mutant with an inactivated *ipt* gene gave the root phenotype in both *Nicotiana* species and in three of the herb species which reacted to this strain.

3.2.2 Use of strains of A. tumefaciens containing the ipt gene under control of its own promoter

Table 4 shows the phenotypes on the two model *Nicotiana* species and the 14 species of aromatic plant species. The *ipt* gene in pBin19 was mated into the disarmed strain LBA 4404 and the phenotypes of LBA 4404 strains with or without the *ipt* gene were compared. No response in any of the 16 species was observed with the disarmed strain. Both *N. tabacum* and, to a smaller

[56] R. Kay, A. Chan, M. Daily, and J. McPherson, *Science*, 1988, **236**, 1299.

Table 2 Phenotypes of transformants following infection of a range of plant species with wild type and 'shooty' mutant strains of Agrobacterium tumefaciens

| | Site of mutation | Plant host | | | | | | | | |
		N. tabacum var. Xanthi	N. rustica	Fennel	Sage	Parsley	Angelica	Mentha spicata	M. piperita	M. citrata
Wild type A348	-	G	G	G	G	G	G	G	G	G
Shooty Mutants of A348[50]										
tms 328:: Tn5	tms1	G/S	G	G	nt	G	G	O	G	G
tms 337:: Tn5	tms1	G/S	G	G/R	nt	G	G	nt	G	G
tms 344:: Tn5	tms1	G/S	G	G	nt	G	O	nt	G	G
tms 352:: Tn5	tms1	G/S	G/S	G	nt	G	O	nt	G	G
tms 355:: Tn5	tms1	G/S	G	G	nt	G	O	G	G	G
tms 369:: Tn5	tms1	G/S	G	G	nt	G	O	G	G	G

G – gall formation, G/S – gall tissue from which shoots develop. G/R – gall tissue developing into roots. nt – not tested. O – no response

Table 3   *Phenotype of transformants following infection of a range of plant species with wild type, 'shooty' and 'rooty' mutants of Agrobacterium tumefaciens*

	Site of mutation	N. tabacum	N. rustica	Plant host Fennel	Sage	Parsley	Angelica	Mentha spicata	M. piperita	M. citrata
Wild type Strain pAch5	-	G	G	G	G	G	G	G	G	G
Shooty mutants[51]										
LBA 4060	tms1	G/S	G	G	G	G	O	G	O	O
LBA 1501	tms2	G/S	G	G	G	G	O	G	G	G
Rooty mutants[51]										
LBA 4210	ipt	G/R	G/R	G/R	G/R	G	O	G	O	G/R

G – gall formation, G/S – gall tissue from which shoots develop. G/R – gall tissue developing into roots, O – no response

Table 4   *Phenotypes of transformants resulting from infection of* Nicotiana Sp. *and various flavour producing species with the* ipt *gene under its endogenous promoter**

Agrobacterium *strain*	LBA4404 (disarmed)	LBA4404/pSSI[52] ipt *gene on endogenous promoter*
Plant host		
Nicotiana tabacum	O	S
N. rustica	O	G/S
Fennel	O	G
Sage	O	G
Parsley	O	G
Basil	O	R
Mentha spicata	O	O
M. piperita	O	O
M. piperita citrata	O	O
M. piperita vulgaris	O	G

O – no response, G – gall formation, S – shoot formation, G/S – galls from which shoots develop, R – root formation.

* No responses to the *ipt* gene construct were observed in Angelica, Dill, Tarragon, Rosemary, Marjoram, Thyme.

extent, *N. rustica* produced shooty teratomas either directly or following an initial disorganized phase when infected by strains bearing the native *ipt* gene. None of the other species responded to the strains by shoot formation; most did not respond at all, a few produced galls, and in one case the unexpected phenotype was found. Thus, basil produced vigorous root formation following transformation with the *ipt* gene; these roots when excised and cleared of bacteria, grew very actively *in vitro* and have been maintained in culture for over a year. These results, taken together with those obtained using the 'shooty' mutants, show that the simple model for shoot induction derived from studies on tobacco is not readily applicable to other plant species. Indeed it suggests that genes other than the three growth regulator biosynthetic genes may influence the phenotype of the transformation and that the endogenous status of the explant material, particularly in relation to the initial hormonal environment, may counteract the effects of the two *tms* and the *ipt* genes.

3.2.3   *Use of nopaline strains of* A. tumefaciens

Table 5 shows the effects of transformation with three nopaline strains of *A. tumefaciens* (T37, C58, and N273) on two test *Nicotiana* species and on 12 herb species. On *N. tabacum*, strain T37 produced galls which had a pronounced tendency to develop into shoots. Of nine inoculations, eight initially produced galls, one third of which went on to develop shoots, and one

Table 5   *Phenotypes of transformants resulting from infection of various* Nicotiana *and flavour producing plant species using nopaline strains of* A. tumefaciens

Plant species	A. tumefaciens *nopaline strains*		
	T37	C58	N273
Nicotiana tabacum	G/S*	G*	G*
N. rustica	G*	G*	nt
Fennel	G/R	nt	nt
Sage	G⁀G	nt	nt
Dill	G⁀G	nt	nt
Parsley	G⁀G	nt	nt
Basil	G/R	nt	nt
Bay	G	nt	nt
Angelica	G	nt	nt
Tarragon	G	nt	nt
Mentha spicata	G⁀G	G⁀G	G⁀G
M. piperita	G⁀G	G⁀G	G⁀G
M. piperita citrata	G⁀S	G⁀S	G⁀S
M. piperita vulgaris	G⁀S	G⁀S	G⁀S

G⁀G – initial gall retaining phenotype after excision, G⁀S – initial gall giving rise to shoots after excision, G/R – initial gall giving rise to roots on explant, G/S – initial gall giving rise to shoots on explant, nt – not tested.

* with low frequency of rooting.

of the nine also developed roots. Neither C58 nor N273 induced shoot formation in tobacco, but between 10 and 30 % of initial galls formed developed roots. A similar situation existed in *N. rustica* with each of the three strains giving a low frequency of root formation but no shoot development. T37 induced gall formation on all of the 12 herb species; in two cases fennel and basil roots were formed spontaneously from the gall tissue. In *Mentha* species, the initial hard galls, which stably maintained this phenotype while attached to the explant, were excised and placed on agar plates containing ampicillin to remove contaminating bacteria. In *M. spicata* and *M. piperita*, the axenic excised galls developed into friable but still undifferentiated galls. However, in both *M. piperita vulgaris* and *M. citrata* the initial hard galls differentiated into shooty teratomas following excision from the explant, decontamination from *Agrobacterium*, and growth on hormone-free media. Similar results were obtained with these two plant species using strains T37, C58, or N273. The shoot cultures of *M. citrata* were slightly different in structure to those of *M. piperita vulgaris* in that the shoots of *M. citrata* were continuously developed from a central core of callus tissue whilst in *M. piperita vulgaris* the culture was more organized, with horizontal stolons which grew into vertical shoots. N273 induced shooty cultures in *M. citrata* which developed as stolon-like shoots, while in *M. piperita vulgaris* the galls when cultured axenically were leafy without stems.

The development of shooty teratomas in two *Mentha* species using wild type nopaline strains and the manner of their development following excision from the explant material poses interesting problems concerning the factors determining the resultant phenotype. A characteristic of nopaline strains is that, in addition to the *ipt* gene which is under the control of a eukaryotic type of promoter, there is an additional cytokinin gene *tzs* present in the Ti plasmid in a region outside the t-DNA close to the virulence region.[57] The *tzs* gene has a high degree of homology with the *ipt* gene and codes for a similar dimethylallyl transferase. It is, however, controlled by a prokaryotic promoter and thus unlike the *ipt* gene is expressed in the bacterium. Since we found that some of our early cultures contained a low level of contamination by bacteria[34] the possibility that expression of the *tzs* gene and bacterial production of cytokinins might influence the phenotype of the plant transformants had to be addressed. We developed a rigorous method for the total elimination of *Agrobacterium* contamination[34] and applied this to all the T37- and C58-derived shoot cultures. In all cases, complete elimination of bacteria was achieved and no significant effects on the phenotype of the shoot cultures were observed. Thus the effect of the *tzs* gene was discounted.

The transformed nature of the T37-derived shoot cultures of both *M. citrata* and *M. piperita vulgaris* was confirmed by demonstrating the presence of nopaline in extracts of the shoot cultures and its absence from extracts of the parent plant tissues. In addition, the presence of T37 t-DNA in the genome of the transformants was confirmed by Southern Blot analysis[34] and by polymerase chain reaction methodology. A Hind III-BamHI digest of the total DNA from T37-derived shoot cultures of *M. citrata* and *M. piperita vulgaris* was separated by agarose gel electrophoresis and analysed by Southern blotting techniques and hybridized with a ^{32}P-labelled probe for the *ipt* gene. All four cultures of each species tested showed the presence of the 720 bp fragment corresponding to the *ipt* gene. A second ^{32}P-labelled probe for the virulence region of the Ti plasmid did not hybridize to the DNA extracted from the axenic shooty teratomas showing that the positive *ipt* signal was not due to the presence of DNA from contaminating *Agrobacterium*. Similar polymerase chain reaction experiments[59] using primers for either the *ipt* gene or part of the virulence region, the virB gene, showed that amplification of the *ipt* gene from target DNA extracted from T37-derived shooty teratomas yielded a fragment of the predicted size (*viz.* 700 bp), while no amplification occurred using the primers for the virB gene. This again demonstrates the transformed nature of the culture and the absence of contaminating *Agrobacterium*.

However, the mechanism by which nopaline strains induce shooty teratoma development in the *Mentha* species studied is unclear. Willmitzer

[57] J.S. Beaty, G.K. Powell, L. Lica, D.A. Regier, E.M.S. MacDonald, N.G. Hommes, and R.O. Morris, *Mol. Gen. Genet.*, 1986, **203**, 274.

[58] A. Spencer, Ph.D. Thesis, University of East Anglia, 1991.

[59] J.D. Hamill, S. Rounsley, A. Spencer, G. Todd and M.J.C. Rhodes, *Plant Cell Rep.*, 1991, **10**, 221.

et al.[46] showed that transformation of tobacco with strain C58 yielded white undifferentiated galls, while similar transformation with strain T37 yielded galls from which occasional shoots developed. Two differences were observed in the levels of t-DNA transcripts present in the two types of transformants. The T37-transformants had a lower level of transcript 5 which is thought to be auxin induced and to be involved in the synthesis of indole lactic acid[60] which itself is believed to interfere with IAA activity. Shooty teratomas induced by T37 in tobacco also had a low, often undetectable, level of the *tms1* transcript perhaps suggesting a decreased capacity for auxin synthesis leading to raised cytokinin/auxin ratios favouring shoot development. There was no difference between T37 and C58 in their phenotypic responses in mint, so an explanation of our findings must await further detailed genetic study. A further complication with *Mentha* is that the ultimate phenotype is only expressed in excised culture not while the transformed tissue is still attached to the parent plant. Here it is possible that there is a major influence of the hormonal environment of the parent plant which overrides the inherent phenotypic character of the transformed tissue.

3.2.4   Use of strains to over-express the ipt gene.

The *ipt* gene was placed under the control of various promoters of increasing strength, namely the nopaline synthase (NOS) promoter, the Cauliflower Mosaic Virus promoter (CaMV35S) and the CaMV35S promoter with duplicated upstream enhancer sequences[57] (E35S) as well as under its endogenous control. In addition a single piece of DNA (Xbal fragment 7) with the *tms1, tms2,* and *ipt* genes (*onc* genes) intact with their own promoters, was cloned into pBIN19 as a control strain. A disarmed strain of *A. tumefaciens* (LBA 4404) was used to house the various *ipt* constructs and the *onc* genes and the resultant strains were tested in transformations with *N. tabacum* and were used to transform both *Mentha citrata* and *Mentha piperita vulgaris.* Table 6 shows the phenotypic responses in these transformations. The *onc* construct induced gall formation in all of these species. All levels of expression of the *ipt* were sufficient to induce rapid shoot teratoma formation in tobacco and these shooty cultures were easily maintained and grew well as *in vitro* cultures after excision from the parent plant. In contrast all the *ipt* constructs induced gall formation in both *Mentha* species. In some transformations with the weaker promoter constructs *i.e.* endogenous-*ipt* on *M. citrata* and NOS-*ipt* on both species, it proved impossible, after clearing them of *Agrobacterium,* to maintain the initial galls in axenic culture. However, in the other cases the initial galls differentiated to produce shooty teratomas following excision and transfer to hormone-free media. This sequence of development parallels the development of shooty teratomas induced by the nopaline strains. Neither of the less powerful promoters is

[60] G. Schell, 'Proc. NATO Advanced Study Institute on Plant Genomes', 1992 (in press).

Table 6 *Phenotype response to various* ipt *and* onc *constructs in* Agrobacterium tumefaciens *LBA 4404 of* N. tabacum *and* Mentha sp.

	N. tabacum	M. citrata		M. piperita vulgaris	
		Initial	*After excision*	*Initial*	*After excision*
Endogenous–*ipt*	S	G	NE	G	G
NOS–*ipt*	S	G	NE	G	NE
CaMV35S–*ipt*	S	G	S	G	S
CaMVE35S–*ipt*	S	G	S	G	S
onc	G	G	G	G	G

S – shooty teratomas directly from explant or after excision, G– gall on explant or after excision, NE– not established in *in vitro* culture.

capable of promoting shoot differentiation and this contrasts with the situation in *Nicotiana*. However both of the more powerful promoters (35S and E35S) induce shoot formation in both species of *Mentha*. These shoot cultures were rather similar in structure; both produced long stems with leaves and had a slight tendency to form roots at the nodes. These cultures were shown to be transformed using polymerase chain reaction methodology[59] demonstrating the presence of the *ipt* gene and the absence of bacterial plasmid genes from outside the t-DNA region indicating the axenic character of the cultures.

On the basis of this work two approaches have proved successful in developing transformed shoot cultures of *Mentha*, namely the use of wild type nopaline strains and the use of strains with the *ipt* under the control of the most powerful plant promoters available. Two T37-derived cultures, one each of *M. citrata* (MC) and *M. piperita vulgaris* (BM1), one 35S-*ipt*- and one E35-*ipt*-derived culture of *M. piperita vulgaris* (BM35S and BME35S respectively) were selected for more detailed study.

3.3 Properties of Transformed Shoot Cultures of *Mentha* sp.: Growth and Monoterpene Production

The T37-derived cultures of both *Mentha* species grew well and maintained their phenotype in culture on simple media. Both Gamborg's B5[61] and MS[62] media-lacking hormones, termed B50 and MS0 respectively, were satisfactory for supporting growth both in their liquid and semi-solid forms. The T37 derived *M. citrata* (MC1) culture grew equally well on both B50 and MS0. Figure 2 shows the growth of MC1 on B50 from an inoculum of 1 g fresh weight to 10 g in about 20 days. *M. citrata* plants accumulate principally the acyclic monoterpenes, linalool and linalyl acetate which comprise over 85 % of the oil. In culture the oil produced reflects the plant in having

[61] O.L. Gamborg, R.A. Miller and K. Ojima, *Expt. Cell Res.*, 1968, **50**, 151.
[62] T. Murishige and F. Skoog, *Physiol. Plant.*, 1962, **15**, 473.

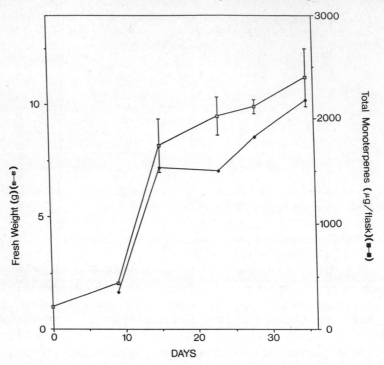

Figure 2   *Growth (□) and total monoterpene accumulation (■) of the T37-derived shoot culture (MC1) of M. citrata grown on B50 medium; 1 g of inoculum was transferred into 50 ml of B50 medium and incubated at 26 °C at 840 lux with a daily cycle of 16 h light and 8 h dark*

over 85 % linalool and linalyl acetate. The linalool/linalyl acetate ratio is constant during the major growth phase but falls late in the growth phase.[58] Figure 2 shows that the production of terpenes follows growth to reach over 2 mg/flask after 35 days.

The T37-derived culture of *M. piperita vulgaris* (BM1) grows from an inoculum of 1 g fresh weight to over 16 g in 40 days (Figure 3). Plants of this species produce a more complex oil that *M.citrata* consisting of cyclic monoterpenes, principally menthol and menthone, with significant levels of minor components such as 1,8-cineol, limonene, piperitone, isomenthone, and menthofuran.[58, 63] The T37-derived culture BM1 produces menthol as the major constituent but accumulates more menthofuran and less menthone than the parent plant. A range of other monoterpenes including menthyl acetate, 1,8-cineol, sabinene, neomenthol, and limonene have also been detected in hexane extracts of BM1. Although menthol is the major monoterpene present during the major growth phase, the level of menthofuran rises steeply at the late growth phase and becomes a major component at this stage.[58] Figure 3 shows that in BM1 monoterpene formation rises to a peak of about 4500 µg/flask at day 20 and then falls.

[63] W.D. Loomis, in 'Biosynthesis and Metabolism of Monoterpenes in Terpenes in Plants', ed. J.B. Pridham, Academic Press, London, 1967, p. 59.

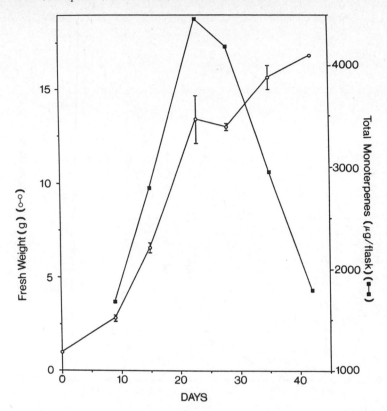

Figure 3 *Growth (○) and total monoterpene accumulation (■) of the T37-derived shoot culture (BM1) of* M. piperita vulgaris *grown on* MS0 *medium. Conditions of cultures as in Figure 2*

Growth and monoterpene production of the *M. piperita vulgaris* cultures derived using the E35S-*ipt* (BME35S) and 35S-*ipt* (BM35S) are compared in Figures 4 and 5. The growth of BME35S is much slower than that of BM35S reaching only 3.5 g fresh weight in 50 days compared with over 8 g in 40 days. The pattern of synthesis of monoterpenes is rather similar in both cultures; however, higher levels of monoterpene are achieved in shorter culture periods in the slower growing culture (BME35S). Both cultures make between 8–10 mg of monoterpene/flask in 50 days of culture. In both cases the synthesis of terpenes follows growth fairly closely, although in BM35S there is a tendency for terpene formation to lag slightly.

The decrease in terpene accumulation observed in culture BM1 beyond 45 days of culture is also found in MC1, BM35S, and BME35S in culture periods beyond 50 days (data not shown). The observed loss of mono-terpenes from the shoot biomass is not simply related to senescence of culture but may be partly due to release of terpenes by outward diffusion from the gland cells or by rupture of the cuticular cover of the gland cells. Small amounts of menthol have been detected in the medium but they do not

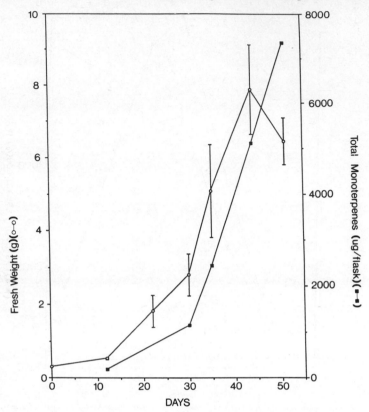

Figure 4   *Growth (○) and total monoterpene accumulation (■) of 35S-ipt derived shoot*
cultures of M. piperita vulgaris. *Conditions of culture as in Figure 2 except that*
the inoculum was 2–300 mg

account for the decreases in total monoterpene content observed. It is possible that metabolism of the released monoterpenes occurs in the medium. It is equally likely that the loss of monoterpenes represents their degradation within the shoot cells at the later stages of culture when the rate of synthesis declines.[24]

It was concluded that these transformed shoot cultures perform well in culture in hormone-free media and produce a range of compounds which reflect those produced by the parent plant. There is a close correlation between the composition of the oil of the plant and of the cultures in the case of *M. citrata*, but in the *M. piperita vulgaris* cultures a significant increase in the relative amount of menthofuran, largely at the expense of menthone, was observed compared with the parent plant. However, it is known that in the plant this ratio is affected by agronomic conditions[63] and the observed differences may represent the different environmental conditions in cultures compared with field-grown plants. The cultures have proved to be stable in monoterpene production over a period and provide ideal systems in which to pursue studies of the biochemical regulation of monoterpene biosynthesis.

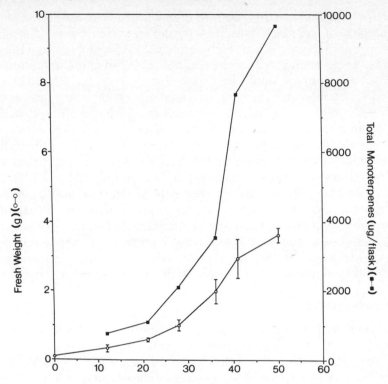

Figure 5   *Growth* (o) *and total monoterpene accumulation* (■) *of an* E35S-ipt *derived shoot culture of* M. piperita vulgaris. *Conditions of culture as in Figure 4*

4   Opportunities for Flavour Improvement by Biotechnology

The availability of tissue culture systems in facilitating detailed biochemical and enzymological studies is an important first step in the attempt to improve flavour yield and quality by genetic manipulation. As we have seen, cell culture systems for simple flavours are available in a few instances and advances such as the use of transformed organ culture suggests that systems producing the more complex flavours such as essential oils will become available in the next few years. With knowledge of the biochemical pathways leading to flavours it may be possible to increase the overall content of flavour compounds in a tissue. Over-expression of key enzymes at the early part of this pathway would be the target for such work. The regulatory enzyme of terpene biosynthesis, hydroxymethylglutaryl CoA reductase (HMG CoA reductase) has been cloned from plant sources[64] and over-expression of this gene might lead to increases in total terpenoid formation,

[64] R.M. Learned and G.R. Fink, *Proc. Nat. Acad. Sci., USA.*, 1989, **86**, 2779.

some of which might be diverted into monoterpenes. An example of success in this type of approach has been in the over-expression of a gene of ornithine decarboxylase, an early enzyme in nicotine biosynthesis in *Nicotiana* roots leading to an increased content of nicotine in the roots.[65] A more subtle approach would be to attempt to divert intermediates out of general terpenoid metabolism by increased expression of enzymes acting at the branch point leading to monoterpenes (*i.e.* monoterpene cyclase).[66]

In addition to increasing total monoterpene yield, another object would be to optimize the production of desirable compounds and to prevent that of undesirable ones using anti-sense RNA technology to block the formation of the enzymes diverting intermediates to the undesirable products. All of these approaches rely on detailed investigation of the enzymology of the system to identifying the opportunities for improvement. Overall significant opportunities will exist in the future to manipulate plant metabolism including that of flavour compounds to desirable ends. Its success is very dependent on defining suitable targets for such work and undertaking the underpinning biochemical and genetic studies necessary to bring it to fruition.

Acknowledgement

The authors wish to acknowledge the contribution of John Reynolds (IFR, Norwich) in the development of the chromatographic methods employed here. They also wish to acknowledge the assistance given by J. Marshall of PFW (UK) Ltd. in the assessment of the essential oils produced in culture and to PFW (UK) Ltd. for their financial support for part of the work related to *Mentha* sp.

[65] J.D. Hamill, R.J. Robins, A.J. Parr, D.M. Evans, J.M. Furze, and M.J.C. Rhodes, *Plant Mol. Biol.*, 1990, **15**, 27.
[66] R. Kjonaas and R. Croteau, *Arch. Biochem. Biophys.*, 1983, **220**, 79.

Flavour Improvement Through Microbial Selection

H.-P. Hanssen

IMB — INDUSTRIELLE MIKROBIOLOGIE UND
BIOTECHNOLOGIE GmbH, LURUP 4, D-2070 GROßHANSDORF,
GERMANY

1 Introduction

Each biotechnological development starts with an intensive study of the relevant literature and an investigation of the patent situation. As a result of an often long and difficult process of reaching a decision, it is concluded that some strains of the micro-organism in question should be ordered from one of the well known culture collections. Then, with the arrival of the cultures from that culture collection, frequently a lot of problems arise and the first laboratory results are often disappointing. The situation described applies to many companies, not only to recently established biotechnology firms, but also to larger companies involved in hitherto traditional fields. It applies to all areas of biotechnological developments, thus also to those involved in the generation of flavour and fragrance chemicals. The intention of this chapter is to give some practical help and advice to beginners in biotechnology, and to show the possibilities, difficulties, and limitations in this field. Since most of our own experiences have been made with filamentous fungal strains, the major part of this chapter will deal with these organisms, but the statements are generally valid and do also apply to other organisms used, like bacteria or yeasts.

Considerations arising in this context comprise:
(a) where and how to get strains and to establish the correct identification of isolates,
(b) the maintenance of cultures in the laboratory,
(c) the performance of screening programmes,
(d) the selection of a suitable strain using defined culture conditions, and
(e) the adaptation and maintenance of the organism to the biotechnological process during 'scale up'.

2   Culture Collections and Alternatives

The role of culture collections as a valuable source for micro-organisms
cannot be underestimated. Quoting filamentous fungi as an illustration,
Table 1 demonstrates the size of some major collections. It is considered
that the number of strains exceeds 170 000, scattered through over 200 collections, and that around 7000 different species are represented.[1] This is an
impressive figure, but, on the other hand, more than 65 000 different
filamentous fungi have been described so far, and it is expected that the
number in nature may well exceed 250 000. This means that only a small percentage of these organisms is currently available from culture collections.

   The existence of such culture collections, with data on the characteristics
of their strains, makes possible enormous savings when appropriate isolates
are available. As pointed out by Krichevsky *et al.*,[2] the best source for strains
will often be from the collection with the most available data, rather than
the most complete selection of isolates. Without diminishing the value of
culture collections, it must be stressed, however, that strains from these
sources may frequently cause a lot of problems. These include wrongly
determined isolates, contaminated cultures, and degenerated strains.

   Thus, the ultimate source for strains with desired properties is isolation
from nature. But this effort can be very labour-intensive with resulting high
costs, especially when no simple screening procedures are available, as in
the identification of isolates producing antibiotics. For flavour and fragrance chemicals, such simple screening systems are not available, apart
from some attempts using TLC procedures. Since the best information
about flavour and fragrance compounds is still obtained with GLC and GC-
MS, simplified methods remain insufficient, resulting frequently in a loss of
further useful particulars. Before discussing the possibilities of efficient
'intelligent' screening programmes, another crucial point, *i.e.* the correct
handling of strains in the laboratory, should be considered.

3   Maintenance of Laboratory Strains

Cultures obtained for one's own collection should be as fully documented as
possible. Hawksworth and Allner[3] have summarized the data a documentation should include (Table 2). Apart from the history of the strain, additional information concerning growth, media, physiological and biochemical
data, as well as indications to pathogenicity should be listed. In particular, if

[1] D.L. Hawksworth, in 'Living Resources for Biotechnology: Filamentous Fungi', ed. D.L.
   Hawksworth and B.E. Kirsop, Cambridge University Press, Cambridge, 1988, p. 1.
[2] M.I. Krichevsky, B.O. Fabricius, and H. Sugawara, in 'Living Resources for Biotechnology:
   Filamentous Fungi', ed. D.L. Hawksworth and B.E. Kirsop, Cambridge University Press,
   Cambridge, 1988, p. 31.
[3] D.L. Hawksworth and K. Allner, in 'Living Resources for Biotechnology: Filamentous
   Fungi', ed. D.L. Hawksworth and B.E. Kirsop, Cambridge University Press, Cambridge, 1988,
   p. 54.

Table 1 *Major culture collections covering fungal strains (modified from Reference 1)*

Culture collection	Country	Number of available strains	Remarks
Agricultural Research Service Culture Collection (NRRL)	USA	44 000	requests may be made by asking for particular taxa; limited to 12 strains
Centraalbureau voor Schimmelcultures (CBS)	NL	29 500	culture sales worldwide; reduction for non-commercial use
American Type Culture Collection (ATCC)	USA	22 000	culture sales worldwide
Mycothèque de l'Université Catholique de Louvain (MUCL)	B	15 000	culture sales worldwide
CAB International Mycological Institute (IMI)	GB	12 500	culture sales worldwide; reduction for non-commerical use
Canadian Collection of Fungus Cultures (CCFC)	CAN	12 000	most isolates available
Institute for Fermentation (IFO)	J	7000	culture sales worldwide
Fungal Genetics Stock Center (FGSC)	USA	5500	Ascomycotina; cultures available for research and teaching
Friedrich-Schiller-Universität Jena (MW)	D	5400	a catalogue is available
Plant Research Division Culture Collection (WA)	AUS	5000	specimens are distributed on request

Table 2 *Documentation of deposited cultures (modified from Reference 3)*

Name, Isolate number	Temperature range for growth
Isolated/Identified by	Optimum growth temperature
Previous history	Light requirements
Other collections in which deposited	Storage method(s)
Host substratum	Physiological properties
Growth substratum	Pathogenicity
Date of isolation	Flavour and fragrance compounds
Dates of subcultures	Other metabolic products
	References to citations in papers/patents

the isolates do not possess the GRAS (generally regarded as safe) status, safety aspects and regulations have to be ensured.

Preservation techniques for micro-organisms have been summarized by Kirsop and Snell[4] and—especially for fungal cultures—by Smith.[5] These methods include:

(*a*) serial transfer on agar,

(*b*) drying, and

(*c*) freezing.

A comparison of these techniques is given in Table 3. Unfortunately, fungi from different taxa may react distinctly differently towards these methods, but in general, freezing and storage in liquid nitrogen at a temperature of −196 °C is the best preservation technique currently available for filamentous fungi. Table 4 shows that the viability using this method is comparatively satisfactory.

Other serious problems, particularly in fungus culture collections, can be formed by mites, since they transfer spores from one culture to the other, causing severe cross-contamination. The best protection against mite invasion is clean work surfaces, the immediate destruction of contaminated cultures by steam, alcohol, or autoclaving, and the storage of cultures by freeze-drying.[5]

4   Culture Conditions Influencing Growth and Metabolic Activities

The major factors affecting growth and metabolic activities are: composition of the culture medium, pH, aeration, temperature, water activity, and, sometimes, light.

Generally, strains of the same species and genera usually grow best on similar media, but growth requirements may vary from strain to strain; the metabolic activity, however, can vary enormously within one species (see below). It should be mentioned that some fungi may deteriorate when kept on the same medium for longer periods, so different media should be alternated from time to time.

The influence of the culture conditions on the metabolic activities can be demonstrated by numerous examples. The effect of different aeration conditons on the bioconversion of 4-methylheptane-3,5-dione has recently been described by Fauve and Veschambre (Figure 1).[6] Only under anaerobic conditons did the yeast-like fungus *Geotrichum candidum* produce the desired metabolite, the insect pheromone sitophilure [(4*R*,5*S*)-(-)-4-methyl-5-hydroxyheptan-3-one].

[4] B.E. Kirsop and J.J.S. Snell, 'Maintenance of Microorganisms', Academic Press, London, 1984.

[5] D.Smith, in 'Living Resources for Biotechnology: Filamentous Fungi', ed. D.L. Hawksworth and B.E. Kirsop, Cambridge University Press, Cambridge, 1988, p. 75.

[6] A. Fauve and H. Veschambre, *Tetrahedron Lett.*, 1987, **28**, 5037.

Table 3 *Comparison of methods of preservation (from Reference 5)*

Method of preservation	Shelf-life	Genetic stability
Serial transfer on agar storage:		
(a) at room temperature	1–6 months	variable
(b) in the refrigerator	6–12 months	variable
(c) under oil	1–32 years	poor
(d) in water	2–5 years	moderate
(e) in the freezer	4–5 years	moderate
Drying:		
(a) soil	5–20 years	moderate to low
(b) silica gel	5–11 years	good
(c) freeze-drying	4–40 years	good
Freezing		
liquid nitrogen storage	infinite	good

Table 4 *Viability of fungi stored in liquid nitrogen (modified from Reference 5)*

Fungal group	Tested	Viable	Viability (%)
1. Mastigomycotina			
Chytridiomycetes	56	9	16
Oomycetes	348	172	50
2. Zygomycotina			
Zygomycetes	267	254	95
3. Ascomycotina			
Clavicipitales	15	13	87
Endomycetales	10	10	100
Ophiostomatales	22	22	100
Sordariales	178	170	96
Other orders			75–100
4. Basidiomycotina			
Hymenomycetes	149	143	96
Gasteromycetes	8	8	100
Uredinomycetes	18	10	56
5. Deuteromycotina			
Hyphomycetes	1543	1465	95
Coelomycetes	238	224	94

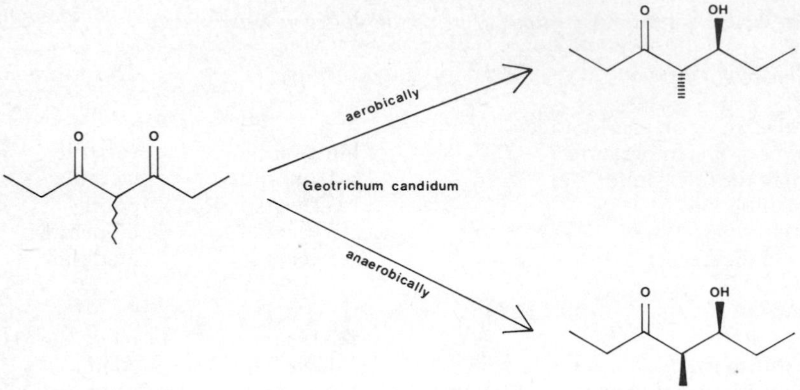

Figure 1   *Stereoselective bioconversion of 4-methylheptane-3,5-dione (from Reference 6)*

Furuhashi[7] has investigated the production of optically active epoxides by different bacteria and yeasts. He demonstrated that the strains grown on glucose did not epoxidize alk-1-enes, whereas they did so when grown on alkanes. An exception to this observation was a strain of the actinomycete *Nocardia corallina*, tested together with 16 other organisms.

We have repeatedly shown the influence of culture conditions on the accumulation of volatiles in fungal cultures. In particular, the nitrogen source affected the yields of certain metabolites distinctly; in individual cases, even metabolic shifts leading to a completely different spectrum of volatiles could be observed. This may be demonstrated with the accumulation of 6-protoilludene in cultures of *Ophiostoma piceae* (Ha 4/82; Table 5), and with the production of cinnamate derivates and sesquiterpenes, respectively, by the basidiomycete *Lentinus lepideus* FPRL 7B.[8]

Ophiostoma piceae is an ascomycete producing mainly the tricyclic sesquiterpene hydrocarbon 6-protoilludene in liquid cultures.[9] Using solid media, we could also isolate two related alcohols, cerapicol and cerpicanol.[10] Testing different inorganic and organic nitrogen sources, the highest yields of 6-protoilludene were obtained when isoleucine and phenylalanine, respectively, were offered.

From fruit-bodies of the basidiomycete *Lentinus lepideus*, several cinnamic acid derivatives have been isolated.[11, 12] Grown in liquid culture media, strains of this species also produce these compounds when, for example, asparagine is offered as sole nitrogen source. Replacing this amino acid by others, like isoleucine, a completely different spectrum of volatiles is

7 K. Furuhashi, *Chem. Econ. Engin. Rev.*, 1986, **18**, 21.
8 H.-P. Hanssen and W.-R. Abraham, in 'Proc.4th European Congress on Biotechnology 1987', Vol. 3, ed. O.M. Neijssel, R.R. van der Meer, and K.Ch.A.M. Luyben, Elsevier, Amsterdam, 1987, p. 291.
9 H.-P. Hanssen, E. Sprecher, and W.-R. Abraham, *Phytochemistry*, 1986 **25**, 1979.
10 H.-P. Hanssen and W.-R. Abraham, *Tetrahedron*, 1988, **44**, 2175.
11 W.B. Turner, 'Fungal Metabolites', Academic Press, London, 1971.
12 W.B. Turner and D.C. Aldridge, 'Fungal Metabolites II', Academic Press, London, 1983.

Table 5   *Effect of nitrogen source on 6-protoilludene accumulation in liquid cultures of*
Ophiostoma piceae Ha 4/82 *(modified from Reference 9)*

N-Source	(%)	N l⁻¹ culture medium (mM)	Maximum of protoilludene accumulation (mg l⁻¹)
Inorganic:			
calcium nitrate	0.10	8.47	0.2
ammonium nitrate	0.05	12.50	1.0
Organic:			
leucine	0.15	11.40	3.6
glutamine	0.10	13.89	4.9
asparagine	0.10	13.30	15.7
alanine	0.10	11.23	17.9
isoleucine	0.15	11.40	34.2
phenylalanine	0.15	9.09	60.6

obtained. Under these culture conditions, almost exclusively sesquiterpene hydrocarbons, alcohols, and ethers are accumulated.[13-15]

The addition of effectors can also affect the generation of a desired product. Thus, Nakamura *et al.*[16] could demonstrate that β-keto esters are reduced stereoselectively into the corresponding L-β-hydroxy esters by baker's yeast, when ethyl chloroacetate was used as an additive.

5   Strain Specificity

The formation of secondary metabolites is frequently a strain-dependent feature. This applies also to flavour and fragrance compounds biosynthesized by various micro-organisms. In the past, we have presented data demonstrating the strain-specific accumulation of terpenes and terpenoids in the genus *Ceratocystis* (Ophiostomatales, Ascomycotina).[17] We have now obtained similar results with fungal strains of the genus *Trichoderma* (Deuteromycotina). This fungus is already successfully used in biopest control. Several non-volatile constituents have been described as biologically active metabolites of *Trichoderma* strains. Besides these, 6-pentyl-α-pyrone (6-PP), a volatile lactone with an intense coconut-like odour, has been identified. About 30 newly isolated strains were screened for their ability to produce low-molecular volatile metabolites. Several strains produced only short-chain alcohols and esters; others produced complex mixtures with sesquiterpenes as predominant constituents. Some of the isolates also produced the desired lactone. We could show that this compound is less sensi-

---

13 H.-P. Hanssen, *Phytochemistry*, 1982, **21**, 1159.
14 H.-P. Hanssen, *Phytochemistry*, 1985, **24**, 1293.
15 W.-R. Abraham, H.-P. Hanssen, and C. Möhringer, *Z. Naturforsch*, 1988, **43C**, 24.
16 K. Nakamura, Y. Kawai, and A. Ohno, *Tetrahedron Lett.*, 1990, **31**, 267.
17 H.-P. Hanssen and E. Sprecher, in 'Flavour '81', ed. P. Schreier, W. de Gruyter, Berlin, 1981, p. 547.

tive against zygomycetes, but highly effective against certain phytopathogenic asco- and deuteromycetes.[18] The yields of 6-pentyl-α-pyrone obtained with different isolates of the species *T. viride*, *T. hamatum*, *T. harzianum*, and *T. koningii* are shown in Figure 2. Using optimized culture conditions, we could increase the production from 35 mg l^{-1} initially to more than 1 g l^{-1}. Compared with strains obtained from different culture collections, our own isolates were generally fitter, with better physiological properties and a distinctly enhanced production of the lactone.

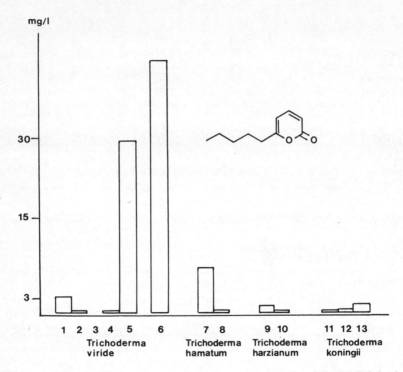

Figure 2 *Strain-dependent formation of 6-pentyl-α-pyrone by different* Trichoderma *isolates*

6 Screening Programmes and Future Prospects

The advantages of using isolates directly from nature are obvious. The exact identification of the isolated material can be achieved by co-operation with experts from universities or other institutions; several culture collections offer a special identification service. So far, screening programmes have been oriented randomly, rather than by intelligent strategies. It can be expected, however, that an increasing knowledge about microbial physiol-

[18] H.-P. Hanssen and I. Urbasch, in '5th European Congress on Biotechnology', ed. C. Christiansen, L. Munck, and J. Villadsen, Munsgaard Intern. Publ., Copenhagen, 1990, p. 372.

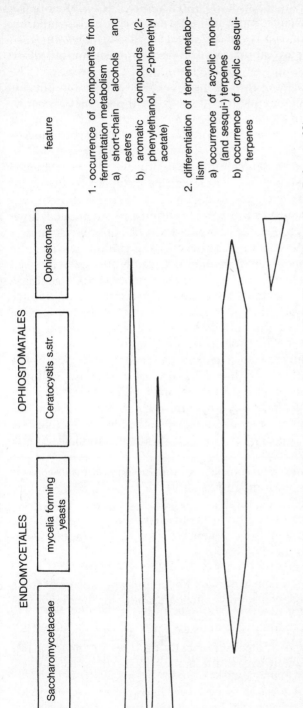

Figure 3 *Comparison of volatiles produced by species of* Endomycetales *and* Ophiostomatales *(from Reference 19)*

ogy, biochemistry, ecology, and genetics will improve this situation. An example for a chemosystematic approach is given in Figure 3, comparing the formation of volatiles in cultures of species of the families *Endomycetales* and *Ophiostomatales* (Ascomycotina). While yeasts produce mainly short-chain alcohols and esters, and — to a certain degree — aromatic compounds, related mycelia-forming species have been described as producers of monoterpenes. This trait can also be observed in the genus *Ceratocystis s. str.*, but seems to be lost in the related *Ophiostoma* group. Strains from this genus have repeatedly been reported as producers of —in part unique — cyclic sesquiterpenes.[19]

Further improvements of a biotechnological process can be achieved using mutants of a promising strain. These can be obtained by treatment with ultraviolet light or with chemical mutagens. For the generation of fungal mutants, frequently fungicides are applied. It must be mentioned, however, that mutants are often less stable compared with the original isolates. A successful process for the production of terpenes using mutants has recently been patented;[20] *Saccharomyces cerevisiae* mutants, which have been blocked in their ergosterol anabolism, produce terpene alcohols like geraniol, farnesol, and linalool. These mutants have been further mutated to produce double mutants, which are additionally defective in alcohol dehydrogenase I (ADH-I) or II. A further mutant, deficient in squalene synthetase activity and ADH-I, was prepared for the production of farnesol.

Further techniques leading to improved strains for the production of flavour and fragrance chemicals include the use of one-spore cultures, protoplast fusion techniques, and molecular–biological methods. The alteration of complex metabolic pathways by molecular–biological methods seems an unrealizable goal in the near future, since the basic principles are only poorly understood.[21] This applies especially to the mechanisms of regulation of such pathways.

In order to make biotechnological processes for the generation of flavour and fragrance chemicals more competitive, efforts in research and development areas should include:

(a) an improved knowledge of microbial metabolism and its regulation,
(b) the optimization of fermentation systems, with special adaptations to the peculiar features of these compounds,
(c) efficient recovery methods,
(d) the search for micro-organisms which can be used as producers of these chemicals, or which possess the enzymatic equipment for the bioconversions desired.

In spite of the current insufficiency of knowledge and techniques, it is foreseeable that a number of biotechnologically derived flavour and fragrance chemicals will find their way to the market in the next decade.

[19] H.-P. Hanssen, 'Proc. Intern. Workshop on Taxonomy and Biology of the Ophiostomatales', Bad Windsheim, August 21–24, 1990, in press.
[20] F. Karst and B.D.V. Vladescu, 1989, Eur. P. 313 465.
[21] G. Mohr, 'Dissertationes Botanicae', Vol. 131, J. Cramer, Berlin, 1989.

Biosynthesis of 3-Isopropyl-2-methoxypyrazine and Other Alkylpyrazines: Widely Distributed Flavour Compounds

E. Leete, J.A. Bjorklund

NATURAL PRODUCTS LABORATORY, DEPARTMENT OF CHEMISTRY, UNIVERSITY OF MINNESOTA, 207 PLEASANT STREET S.E., ST. PAUL, MINNESOTA 55455, USA

G.A. Reineccius, T.-B. Cheng

DEPARTMENT OF FOOD SCIENCE AND NUTRITION, UNIVERSITY OF MINNESOTA, ST. PAUL, MINNESOTA 55108, USA

1 Introduction

Pyrazines (1,4-diazines) are widespread in nature. However, very little definitive information is available on their origin. Some are undoubtedly of biosynthetic origin, produced by enzymatic reactions which may be unique to the species in which the pyrazines are found. Others may be formed by non-enzymatic reactions between molecules which are present in a plant or other species. Some found in cooked or roasted foods are probably produced by non-enzymatic reactions between potential precursors found in the fresh foods or are formed during the heating process. In this chapter we will review the various hypothetical routes to the pyrazines and describe our own work on the biosynthesis of 3-isopropyl-2-methoxypyrazine in the microbial system, *Pseudomonas perolens*.[1] The low molecular weight pyrazines have a pronounced odour which is partially responsible for the characteristic flavours of many foods, both fresh and cooked materials.

---

[1] T.-B. Cheng, G.A. Reineccius, J.A. Bjorklund, and E. Leete, *J. Agric. Food Chem.*, 1991, **39**, 1009.

2   The Occurrence of Pyrazines

Pyrazines have been found in insects and their function in these species may be as alarm or trail pheromones. Some of these pyrazines are illustrated in Figure 1.[2-4] No work has been reported on the origin of these pyrazines in the insects. However, the presence of 3-isopropyl-2-methoxypyrazine in the beetle *Metriorrhynchus rhipidius*,[3] many plants, and some micro-organisms, may indicate that this pyrazine has the same biosynthetic origin in all these different species. The presence of the isopentyl[2, 4] and citronellyl[4] side-chain in some of these insect pyrazines suggests that they are partially formed from terpenoid precursors. The amount of pyrazines found in these insects is quite small; about 1 ng of the mixture of pyrazines being present in each *M. rhipidius* beetle.[3]

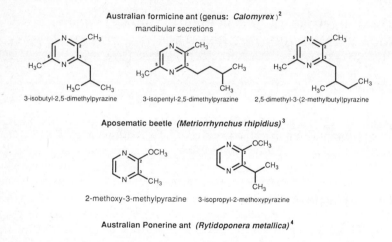

Figure 1   *Pyrazines in insects*

Pyrazines have been found in several micro-organisms, giving rise to characteristic odours of these organisms. Figure 2 illustrates some of these microbial pyrazines.[5-10] Some of these organisms produce quite large

[2] W.V. Brown and B.P. Moore, *Insect Biochem.*, 1979, **9**, 451.
[3] B.P. Moore and W.V. Brown, *Insect Biochem.*, 1981, **11**, 493.
[4] J.J. Brophy, G.W.K. Cavill, and W.D. Plant, *Insect Biochem.*, 1981, **11**, 307.
[5] T. Kosuge and H. Kamiya, *Nature*, 1962, **193**, 776.
[6] A. Gallois and P.A.D. Grimont, *Appl. Envir. Microbiol.*, 1985, **50**,1048.
[7] A.L. Demain, M. Jackson, and N.R. Trenner, *J. Bacteriol.*, 1967, **94**, 323.
[8] A. Miller, R.A. Scanlan, J.S. Lee, L.M. Libbey, and M.E. Morgan, *Appl. Microbiol.*, 1973, **25**, 257.

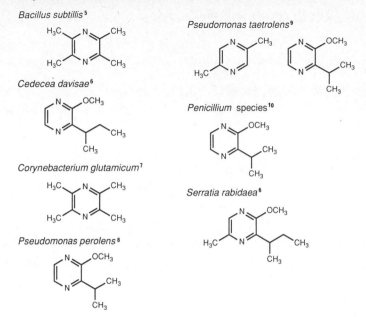

Figure 2 *Pyrazines in micro-organisms*

amounts of the pyrazines; for example, 3 g of tetramethylpyrazine per litre of culture medium was produced by *Corynebacterium glutamicum*.[7]

Many pyrazines have been found in cooked foods, and a review by Maga[11] lists the vast number of alkylpyrazines and related compounds which have been detected in various foods. Some representative examples illustrate the diversity of compounds which have been found in roasted almonds (Figure 3),[12,13] grilled beef (Figure 4),[14,15] and beer (Figure 5).[16] Other foods which have been shown to contain a multitude of alkylpyrazines include chocolate, cocoa, coffee, heated eggs, roasted filberts, cooked Macademia nuts, baked potatoes, cooked rice, black tea, and green tea.[11] It is significant that many more pyrazines were found in the baked potato than were found in the uncooked potato. This almost certainly indicates that the pyrazines found in the baked potato are the products of non-enzymatic reactions. Pyrazines have also been found in tobacco smoke.[17-19]

9   J.H. Meilinger, E.H. Marth, R.C. Lindsay, and D.B. Lund, *J. Food Prot.*, 1982, **45**, 1098.
10   C. Karabadian, D.B. Josephson, and R.C. Lindsay, *J. Argic. Food Chem.*, 1985, **33**, 339.
11   J.A. Maga, *C.R.C. Critical Rev. in Food Sci. and Nutr.*, 1982, **16**, 1.
12   Y. Takei, K. Shimada, S. Watanabe, and T. Yamanishi, *Agric. Biol. Chem.*, 1974, **38**, 645.
13   Y. Takei and T. Yamanishi, *Agric. Biol. Chem.*, 1974, **38**, 2329.
14   I. Flament and G. Ohloff, *Helv. Chim. Acta*, 1971, **54**, 1911.
15   I. Flament, M. Kohler, and R. Aschiero, *Helv. Chim. Acta*, 1976, **59**, 2308.
16   R.J. Harding, H.E. Nursten and J.J. Wren, *J. Sci. Food Agric.*, 1977, **28**, 225.
17   E. Demole and D.A. Berthet, *Helv. Chim. Acta*, 1972, **55**, 1866.
18   D.L. Roberts and W.A. Rohde, *Tobacco Sci.*, 1972, **16**, 107.
19   R.A. Lloyd, C.W. Miller, D.L. Roberts, S.A. Giles, J.P. Dickerson, N.H. Nelson, C.E. Rix, and P.H. Ayers, *Tobacco Sci.*, 1976, **20**, 43.

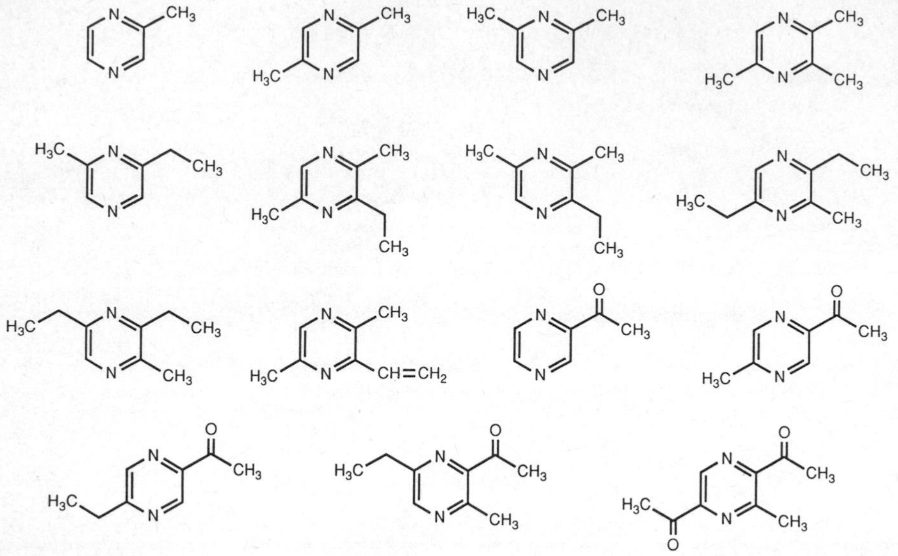

Figure 3 *Pyrazines in roasted almonds*

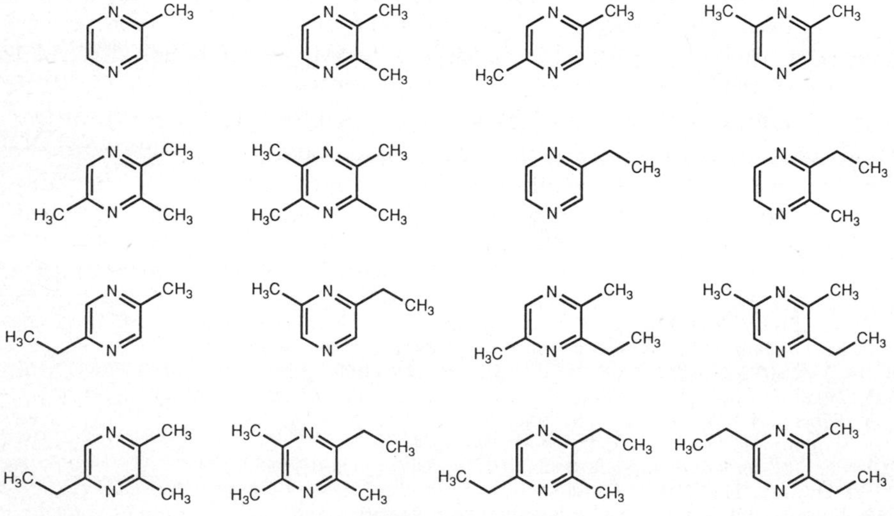

Plus 15 more complex pyrazines

Figure 4 *Pyrazines in grilled beef*

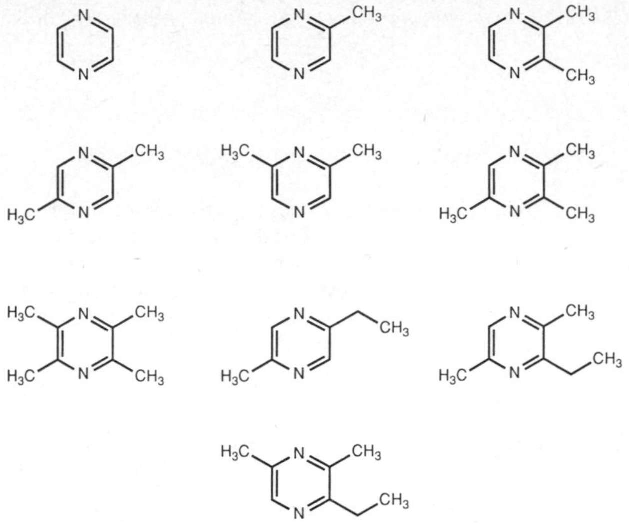

Figure 5 *Pyrazines in beer*

Pyrazines have been found in many raw vegetables and other plants.[20, 21] Table 1 lists the relative amounts of the common 3-alkyl-2-methoxy-pyrazines which were detected in the headspace above the juice squeezed from some common vegetables.[21] The amounts present vary enormously. In fresh peas, 3-isopropyl-2-methoxypyrazine is present at a concentration of one part in 10^{11} parts of the peas.[20]

3 Previous Work on the Origin of the Pyrazines

Aspergillic acid, illustrated in Figure 6, is an *N*-hydroxypyrazin-2-one. Its biosynthesis has been studied in *Aspergillus flavus*.[22] Its precursors are L-isoleucine and L-leucine. It was proposed that these two amino acids condense to yield the 2,5-dioxopiperazine (*cyclo*-L-leucine-L-isoleucine) which is then converted to aspergillic acid via deoxyaspergillic acid. This latter compound is indeed an excellent precursor of aspergillic acid.[22] The sec-butyl side-chain of aspergillic acid has the same (*S*) configuration as that found in L-isoleucine, a result consistent with the direct incorporation of this amino acid. The 2,5-dioxopiperazine was also found to be a precursor of aspergillic acid;[22] however, it was a less efficient precursor than the individual amino acids, indicating that it undergoes hydrolysis to the individual amino acids prior to incorporation. In a review on the 2,5-dioxopiperazines, Sammes[23] was of the opinion that the *cyclo*-L-leucine-L-isoleucine is not a direct precursor of aspergillic acid.

[20] K.E. Murray, J. Shipton, and F.B. Whitfield, *Chem. Ind.*, 1970, 897.
[21] K.E. Murray and F.B. Whitfield, *J. Sci. Food Agric.*, 1975, **26**, 973.
[22] J.C. MacDonald, *J. Biol. Chem.*, 1961, **236**, 512.
[23] P.G. Sammes, *Prog. Chem. Org. Nat. Prod.*, 1975, **32**, 51.

Table 1   *3-Alkyl-2-methoxypyrazines detected in some common vegetables*

Common name (*part extracted*)	Botanical name	3-isopropyl-2-methoxy-pyrazine	3-sec-butyl-2-methoxy-pyrazine	3-isobutyl-2-methoxy-pyrazine
		Amount of pyrazine, in ng, found in the headspace above 500 ml of fresh juice saturated with NaCl		
Asparagus (spears)	*Asparagus officinalis*	30	<5	none
French beans (whole pod)	*Phaseolus vulgaris*	50	5	120
Beetroot (roots)	*Beta vulgaris*	20	500	50
Carrot (root)	*Daucus carota sativa*	<10	250	none
Cucumber (whole)	*Cucumis sativis*	20	<2	none
Lettuce (leaves)	*Latuca sativa*	110	45	10
Nasturtium (whole plant)	*Tropaeolum majus*	300	250	500
Parsnip (root)	*Pastinaca sativa*	<5	550	<5
Garden peas	*Pisum sativum*			
(pod)		1400	20	2
(pea seed)		4	<1	trace
Sweet peppers (bell and green peppers, fruit)	*Capsicum annuum* var. *grossum*	200	300	20,000
Red pepper (fruit)	*Capsicum fructescens*	110	15	5500
Potato (tuber)	*Solanum tuberosum*	10	none	3
Sweet corn (kernels)	*Zea mays* var. *saccharata*	trace	none	none

Jones[24] discovered a facile chemical synthesis of 2-hydroxypyrazines from α-dicarbonyl compounds and the amides of α-amino acids, as depicted in Figure 7. The reaction occurs under very mild conditions in water or aqueous methanol, and some examples of the pyrazines which have been prepared by this method are shown in Figure 8. This reaction led Murray and co-workers[20,21] to propose that the 3-alkyl-2-methoxypyrazines, illustrated in Table 1, are formed by this mechanism; the isopropyl, isobutyl, and sec-

---

[24] R.G. Jones, *J. Am. Chem. Soc.*, 1949, **71**, 78.

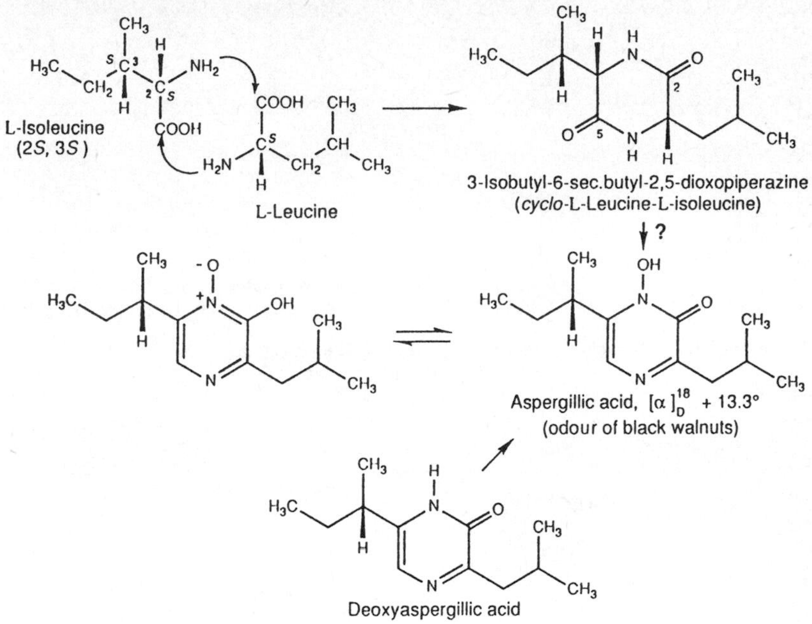

Figure 6 *Biosynthesis of aspergillic acid*

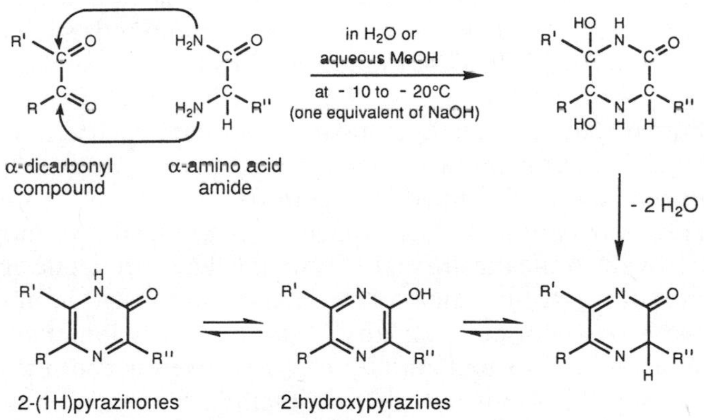

Figure 7 *Jones' synthesis of hydroxypyrazines*

butyl side-chains being derived from valine, leucine, and isoleucine, respectively. This hypothesis was challenged,[25] since it was pointed out that the amides of the common α-amino acids are not found in nature, nor are the α-dicarbonyl compounds required for Jones' reaction. We agree with this opinion as regards the amides of the common α-amino acids. Indeed, few amides of any α-amino acids have been found; one rare example being the

[25] H.E. Nursten and M.R. Sheen, *J. Sci. Food Agric.*, 1974, **25**, 643.

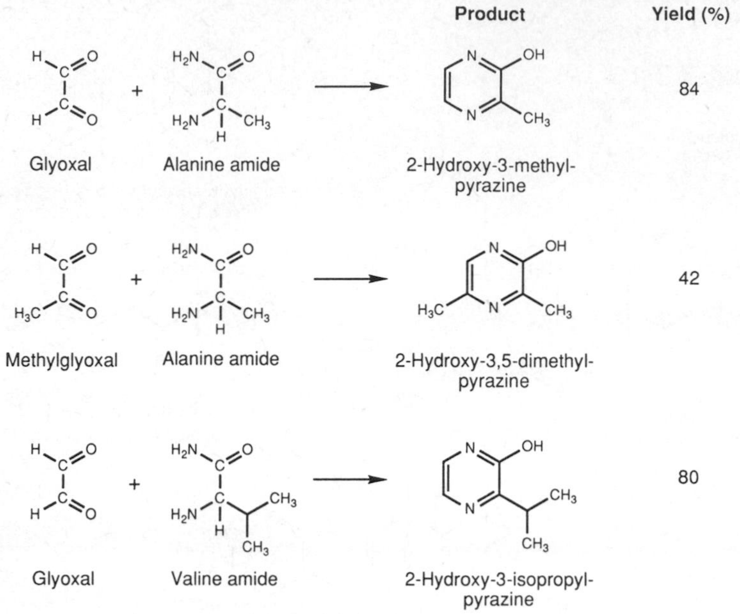

Figure 8 *Some examples of Jones' synthesis*

antibiotic bleomycin, which is produced by *Streptomyces vesticillus*,[26] and is depicted in Figure 9. The low molecular weight compounds such as glyoxal and methylglyoxal are difficult to isolate, but they have been detected in several natural systems, including cheese, by Griffith and Hammond.[27] These authors showed that 2,5-dimethylpyrazine is formed by reaction of lysine with 1,3-dihydroxyacetone in a non-enzymatic system. A possible mechanism for this reaction is illustrated in Figure 10.

Acetoin (3-hydroxybutanone) is formed in nature from two molecules of pyruvic acid by the route illustrated in Figure 11. The acetolactic acid, which is a precursor of valine, undergoes decarboxylation to yield acetoin. Rizzi[28] has obtained good yields of tetramethylpyrazine by reaction of acetoin with ammonium acetate in a small amount of water. Even at room temperature (22 °C), a 13 % yield of this pyrazine is obtained. The yield increases at a higher reaction temperature. A plausible reaction mechanism is illustrated in Figure 12. It is proposed that reaction of acetoin with ammonia under slightly acidic conditions yields 3-aminobutanone which then undergoes self condensation yielding tetramethyl-1,4-dihydropyrazine. Oxidation then affords the tetramethylpyrazine. Bubbling air through the reaction mixture did not increase the yield of the pyrazine, and Rizzi suggested that the oxidation of the intermediate dihydropyrazine occurs by a disproportion-

[26] T. Fukuoka, Y. Muraoka, F. Akio, H. Naganawa, T. Hiroshi, T. Takita, and H. Umezawa, *J. Antibiotics*, 1980, **33**, 114.
[27] R. Griffith and E.G. Hammond, *J. Dairy Sci.*, 1989, **72**, 604.
[28] G.P. Rizzi, *J. Agric. Food Chem.*, 1988, **36**, 349.

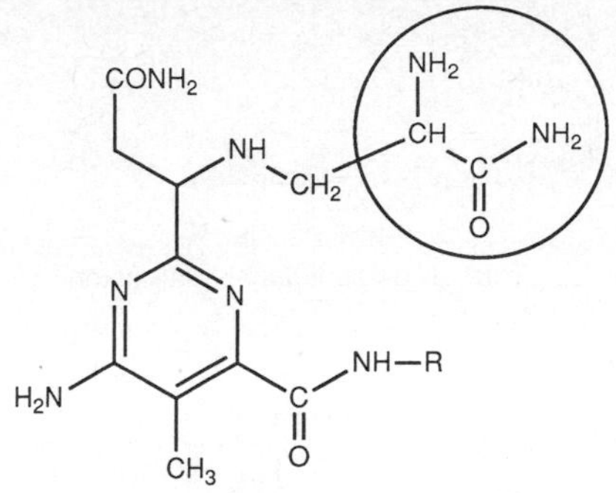

Figure 9 *Bleomycin — a rare amide of an α-amino acid*

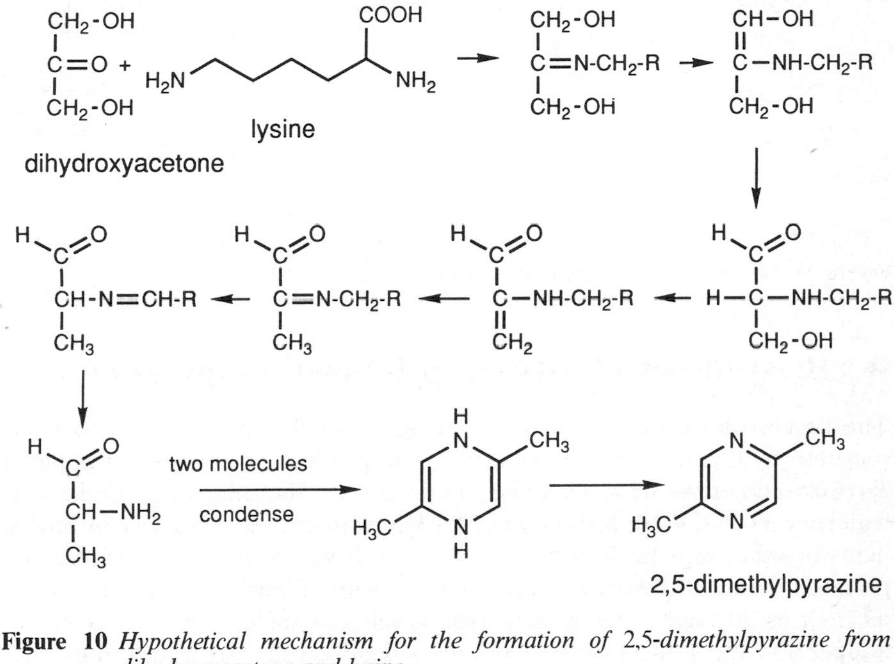

Figure 10 *Hypothetical mechanism for the formation of 2,5-dimethylpyrazine from dihydroxyacetone and lysine*

ation reaction. 1,4-Dihydropyrazines are also plausible precursors of the insect pyrazines which contain a terpenoid side chain. Figure 13 illustrates a hypothetical route to such compounds. Electrophilic attack by the dimethylallyl carbocation generated from dimethylallyl pyrophosphate on the enamine system in 2,5-dimethyl-1,4-dihydropyrazine yields 3-isopentyl-2,5-dimethylpyrazine after oxidation of the ring and reduction of the side-chain double bond.

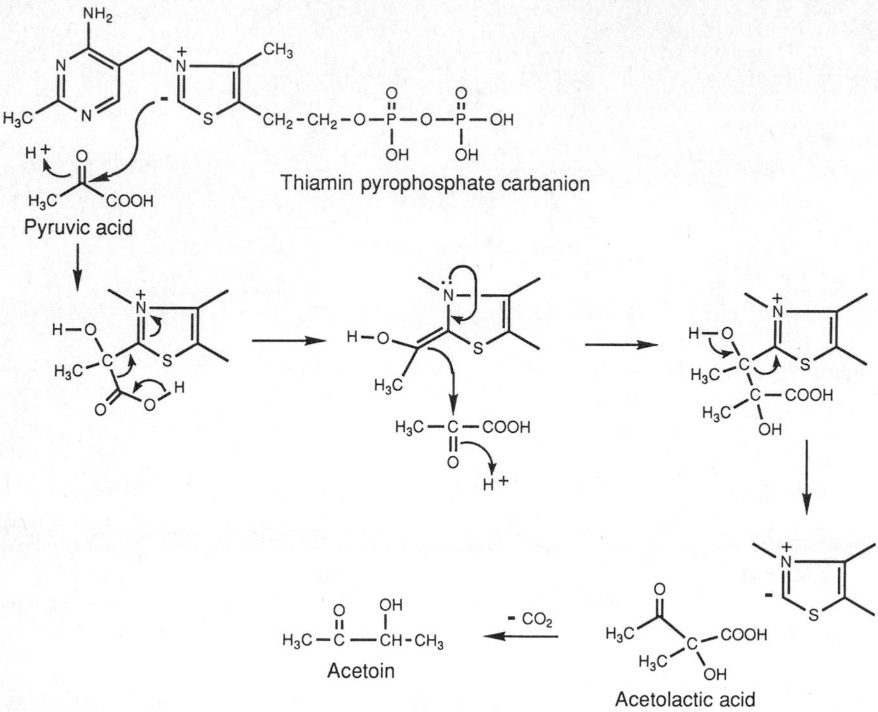

Figure 11 *The metabolism of pyruvic acid to acetoin*

4 Biosynthesis of 3-Isopropyl-2-methoxypyrazine

The first work on the biosynthesis of 3-isopropyl-2-methoxypyrazine was reported by Gallois and co-workers.[29] The production of this pyrazine by *Pseudomonas taetrolens* was studied in a medium containing (g l^{-1} of distilled water) NaH$_2$PO$_4$ (1.5), KH$_2$PO$_4$ (1.0), MgSO$_4$ (0.2), plus a variety of amino acids or other organic compounds (1 g l^{-1}). It was found that L-valine and glycine stimulated the production of 3-isopropyl-2-methoxypyrazine; levels as high as 10 mg l^{-1} being obtained. When glyoxal or glyoxylic acid was added, no production of this pyrazine was observed. The addition of L-[1-^{13}C]valine (1 g l^{-1}) yielded the pyrazine labelled with one atom of ^{13}C, which was detected by mass spectrometry. The molecular ion (*m/z* 152 in the unenriched pyrazine) now appeared at *m/z* 153. The prominent parent ion at *m/z* 137 (due to loss of the methyl group attached to oxygen) appeared at *m/z* 138. The results indicated an almost 100 % incorporation of carbon 1 of valine into the pyrazine. The ^{13}C NMR spectrum of the enriched material showed a large enhancement of the signal at 158.6 p.p.m. which is the one assigned to C-2. However, when [1-^{13}C]glycine was added to the culture

[29] A. Gallois, A. Kergomard, and J. Adda, *Food Chem.*, 1988, **28**, 299.

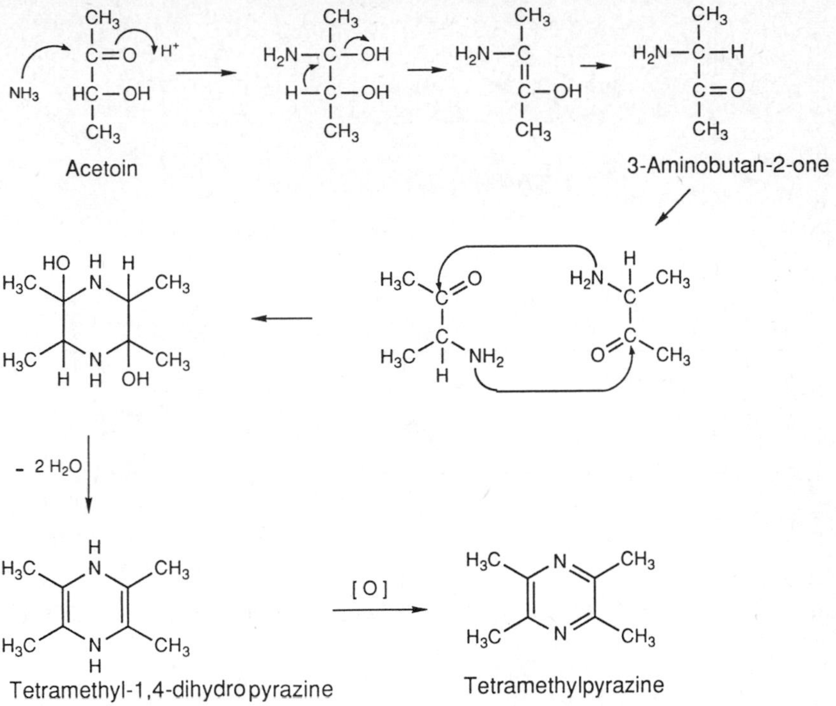

Figure 12 *Reaction of acetoin with ammonia to yield tetramethylpyrazine*

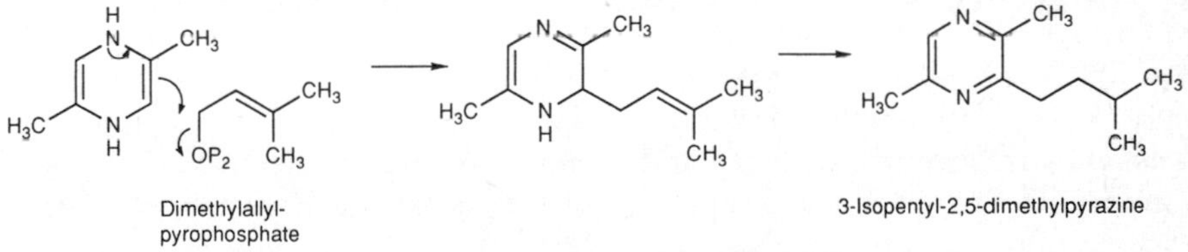

Figure 13 *Hypothetical formation of the insect pyrazines containing a terpenoid side-chain*

medium, along with unlabelled L-valine, no enrichment of the isolated pyrazine was observed. This result would seem to invalidate the proposal that the pyrazine is formed via the 3-isopropyl-2,5-dioxopiperazine as shown in Figure 14.

Independently, we have investigated the biosynthesis of the same pyrazine in *Pseudomonas perolens* grown on a very similar medium to that used by Gallois *et al.*[29] Our experimental set-up is shown in Figure 15. The culture flask was shaken and, every 24 hours, filtered air flushed the vapours above the culture into a trap containing 6 N HCl for one hour. As the cells multiplied, the pH of the solution increased from 6.8 to almost 9. The cells were grown on a medium which contained sodium [2-^{13}C]- or [3-^{13}C]-

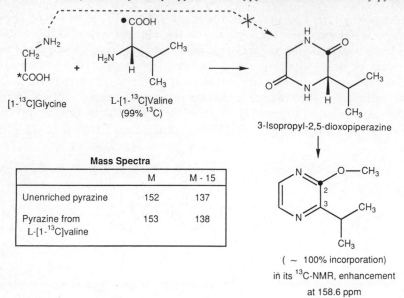

Mass Spectra		
	M	M - 15
Unenriched pyrazine	152	137
Pyrazine from L-[1-^{13}C]valine	153	138

3-Isopropyl-2,5-dioxopiperazine

(~ 100% incorporation) in its ^{13}C-NMR, enhancement at 158.6 ppm

Figure 14 *Biosynthesis of 3-isopropyl-2-methoxypyrazine in* Pseudomonas taetrolens

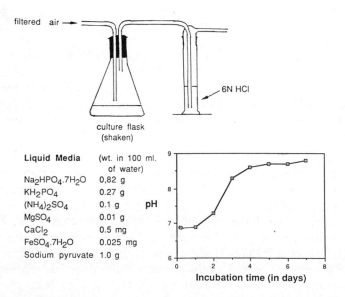

Liquid Media	(wt. in 100 ml. of water)
Na$_2$HPO$_4$.7H$_2$O	0,82 g
KH$_2$PO$_4$	0.27 g
(NH$_4$)$_2$SO$_4$	0.1 g
MgSO$_4$	0.01 g
CaCl$_2$	0.5 mg
FeSO$_4$.7H$_2$O	0.025 mg
Sodium pyruvate	1.0 g

Figure 15 *Culture of* Pseudomonas perolens *in a medium containing sodium pyruvate as a sole source of carbon*

pyruvate (10 g l^{-1}). The pyrazine hydrochloride (~ 0.1 mg in each experiment) was obtained by evaporation of the 6 N HCl trap. Our working hypothesis was the same as that of Gallois, that the pyrazine is formed from the 3-isopropyl-2,5-dioxopiperazine derived from valine and glycine. It is, thus, necessary to consider the established metabolism of pyruvic acid to valine and glycine. Acetolactic acid is formed from two molecules of pyruvic acid (as illustrated in Figure 11) and will be labelled as indicated in Figure 16. The valine derived from [2-^{13}C]pyruvic acid is thus labelled at its C-2 and

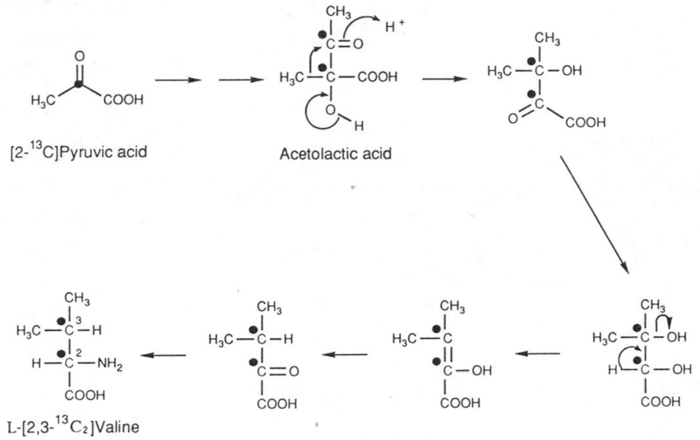

Figure 16 *Metabolism of [2-^{13}C]pyruvic acid to [2,3-^{13}C$_2$]valine*

C-3 positions. [1-^{13}C]Glycine is formed from [2-^{13}C]pyruvic acid by the route illustrated in Figure 17. An oxidative decarboxylation of the [2-^{13}C]pyruvate affords [1-^{13}C]acetyl CoA. When this enters the Krebs cycle, all the carboxylic groups present in the acids of the cycle become labelled after one turn of the cycle. Cleavage of isocitric acid (the glyoxylate shunt) then yields [1-^{13}C]glyoxylic acid which on transamination yields [1-^{13}C]glycine. Figure 18 illustrates the metabolism of [3-^{13}C]pyruvic acid to afford [4,5-^{13}C$_2$]valine. Entry of [3-^{13}C]pyruvic acid into the Krebs cycle intermediates, via [2-^{13}C]acetyl CoA, is more complicated since some of the carboxyl groups become labelled after several turns of the cycle (Figure 19). The distribution of label has been discussed in detail.[30, 31] Thus, the glycine and glyoxylic acid become labelled on both carbons. At steady state conditions, the ratio of label at C-1 and C-2 of these compounds is 1:2.

The labelled 3-isopropyl-2-methoxypyrazine obtained from the pyruvate feedings was examined by ^{13}C NMR spectroscopy and by mass spectrometry. In order to draw conclusions from the ^{13}C NMR spectra, it was important to obtain unequivocal assignments for the signals in the ^{13}C NMR spectrum of

[30] I.D. Spenser, 'Comprehensive Biochemistry Vol. 20', ed. M. Florkin and E.H. Stotz, Elsevier, Amsterdam, 1988, p. 231.
[31] E.Leete, *Acc. Chem. Res.*, 1971, **4**, 100.

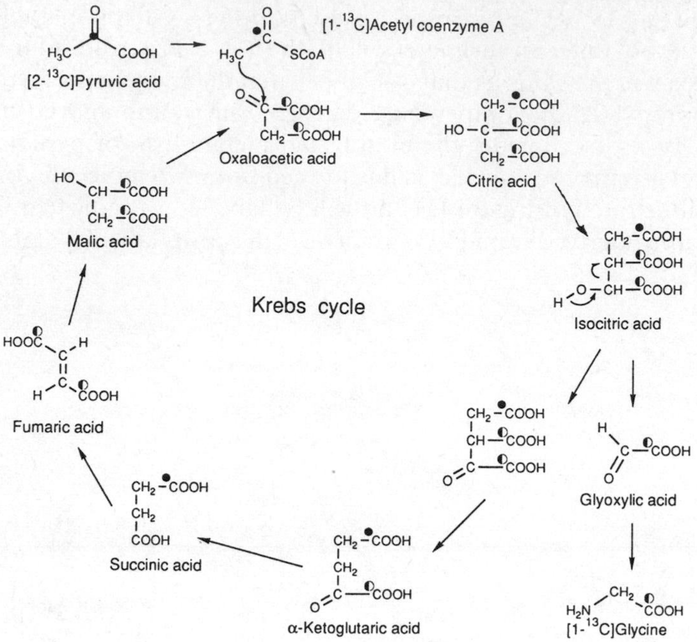

Figure 17 *Metabolism of [2-¹³C]pyruvic acid to [1-¹³C]glycine*

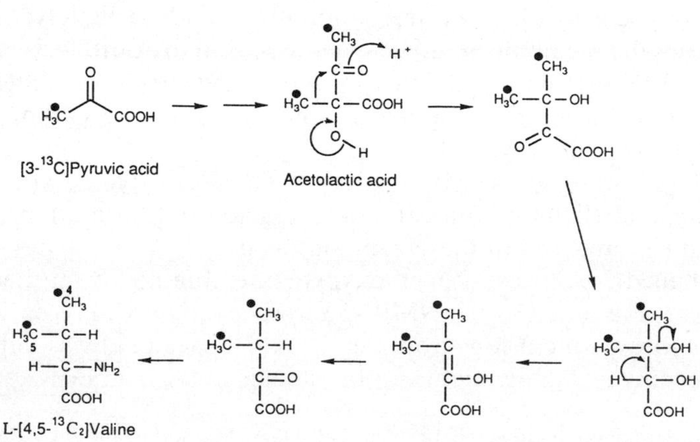

Figure 18 *Metabolism of [3-¹³C]pyruvic acid to [4,5-¹³C₂]valine*

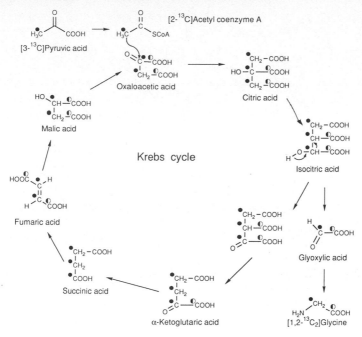

Figure 19 *Metabolism of* [3-¹³C]*pyruvic acid to* [1,2-¹³C₂]*glycine*

the pyrazine, which was determined as its hydrochloride salt in D_2O. The chemical shifts were found to be quite sensitive to changes in pH. The ¹H NMR spectrum of 3-isopropyl-2-methoxypyrazine has been studied in detail.[32,33] A two-dimensional HETCOR pulse sequence between the ¹³C and ¹H spectra afforded a definitive assignment of the ¹³C NMR signals. The ¹³C NMR of the pyrazine derived from [2-¹³C]pyruvate is illustrated in Figure 20. The signals arising from C-3 and C-7 appear as doublets, which are due to the presence of contiguous ¹³C atoms in the enriched pyrazine, resulting in spin-spin coupling. The observed coupling constant (47.3 Hz) is typical for sp^2-sp^3 coupled carbons.[34] This result is consistent with the formation of [2,3-¹³C₂]valine from [2-¹³C]pyruvate, as previously discussed. Complimentary results were observed in the pyrazine derived from [3-¹³C] pyruvate (Figure 21). Here, a large enhancement of the signal due to the chemically equivalent C-8 and C-9 is observed. The signals for C-5 and C-6 exhibit satellites with a coupling constant of 57.2 Hz, which is the typical coupling observed for aromatic carbons.[34] This result indicates that [3-¹³C]pyruvate is affording a two carbon compound labelled with two ¹³C atoms, which ultimately become C-5 and C-6 of the pyrazine. The relative intensity of the C-5 and C-6 signals in the spectra illustrated in Figures 20 and 21 are not consistent with this two carbon compound being glycine.

[32] A.F. Bromwell and R.D. Wells, *Tetrahedron*, 1972, **28**, 4155.
[33] A.F. Bromwell and R.D. Wells, *Tetrahedron*, 1973, **29**, 3939.
[34] V. Wray, *Progress in NMR Spectroscopy*, 1979, **13**, 177.

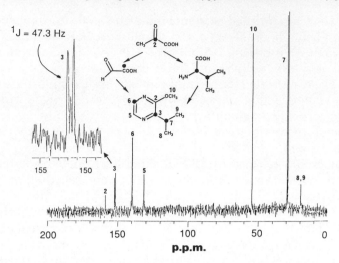

Figure 20 ^{13}C *NMR spectrum of 3-isopropyl-2-methoxypyrazine derived from* [2-^{13}C] *pyruvic acid determined in* D$_2$O *as its hydrochloride salt* (~ 0.1 mg *in* 0.4 ml D$_2$O)

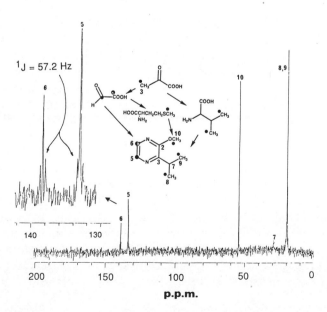

Figure 21 ^{13}C *NMR spectrum of 3-isopropyl-2-methoxypyrazine derived from* [3-^{13}C]*pyruvic acid*

Thus, if [1-13C]glycine had been incorporated via the dioxopiperazine illustrated in Figure 14, the signal at C-5 should have been enhanced relative to C-6. The reverse is observed. Similarly the [1,2-13C₂]glycine with more label at C-2 should have resulted in a larger enhancement of C-6 relative to C-5 (Figure 21). A possible rationalization of our results is illustrated in Figure 22. It is proposed that the two carbon unit which reacts with valine is glyoxylic acid. The resultant Schiff base **1** could then undergo a reduction and reaction with ammonia to yield **2**. Jones' reaction then yields the 2-hydroxy-3-isopropylpyrazine, as previously described (Figure 7). An alternate route would be reacton of **1** with ammonia to yield the imide **4**. Partial reduction of the imide affords **3**, which on dehydration yields the hydroxypyrazine. One problem with this hypothesis is the observation of Gallois,[29] that glyoxylic acid inhibited the production of the pyrazine when it was added to the culture medium. The final step in the scheme depicted in Figure 22 is *O*-methylation with *S*-adenosyl-L-methionine. Rizzi[35] has obtained small yields (0.2–0.5 %) of 3-isopropyl-2-methoxypyrazine by heating the 2-hydroxy-3-isopropylpyrazine with a variety of natural compounds which contain methyl groups (pectin, betaine, choline chloride, and trigonelline). However, we consider that this final methylation is an enzymatic reaction.

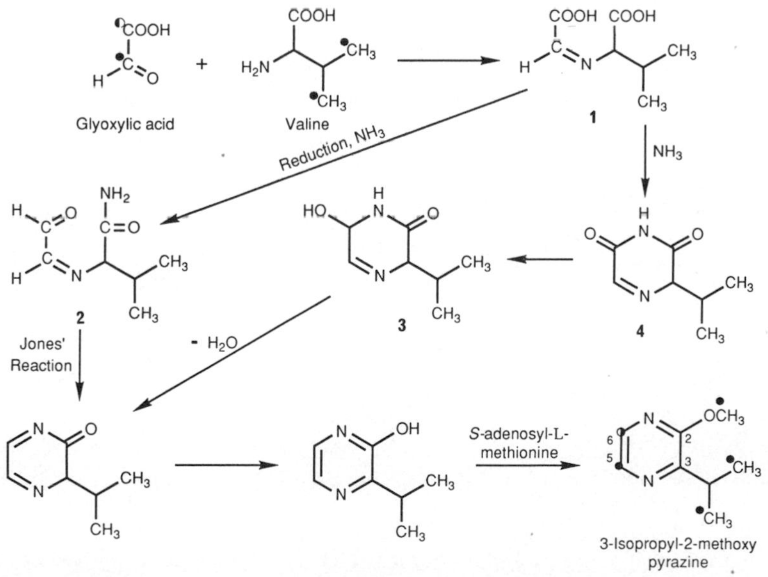

Figure 22 *Hypothesis for the biosynthesis of 3-isopropyl-2-methoxypyrazine consistent with the 13C NMR data*

35 G.P. Rizzi, *J. Agric. Food Chem.*, 1990, **38**, 1941.

The pyrazine derived from [3-^{13}C]pyruvate exhibited considerable enhancement of the signal arising from the *O*-methyl group (C-10). This can be rationalized by the metabolic pathway depicted in Figure 23. The glyoxylic acid reacts with tetrahydrofolic acid ultimately affording the methylfolic acid. The *N*-methyl group is then transferred to homocysteine to yield methionine. The methyl group of methionine is thus derived from the formyl group of glyoxylic acid which is derived from the methyl group of acetic and pyruvic acid. Few have commented on this source of *O*- and *N*-methyl groups in natural products. An example of this metabolic pathway is illustrated in Figure 24. Sulochrin is a polyketide derived from nine acetate units. Thus, the administration of [2-^{14}C]acetate yielded labelled sulochrin, and a partial degradation yielded data consistent with the pattern of labelling illustrated in Figure 24.[36] It should be noted that a small but significant (2.4 %) activity was located on the *O*-methyl group. Another example involving a higher plant is illustrated in Figure 25.[37] [2-^{14}C]Acetate afforded nicotine which was labelled, as illustrated, in the five-membered pyrrolidine ring and on its *N*-methyl group. As indicated in Figure 19, methyl labelled acetic acid affords α-ketoglutaric acid, which is converted to ornithine with the same pattern of labelling. Decarboxylation of the ornithine yields putrescine, an established precursor of the pyrrolidine ring. Here also a significant amount of activity was found on the *N*-methyl group.

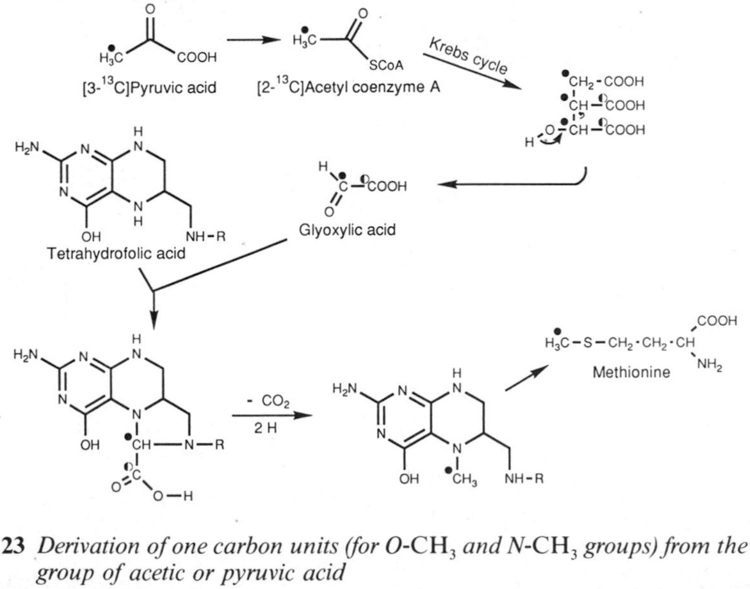

Figure 23 *Derivation of one carbon units (for O-CH$_3$ and N-CH$_3$ groups) from the methyl group of acetic or pyruvic acid*

[36] R.F. Curtis, P.C. Harries, C.H. Hassall, J.D. Levi, and D.M. Phillips, *J. Chem. Soc. (C)*, 1966, 168.
[37] T. Griffith and R.U. Byerrum, *Science*, 1959, **129**, 1485.

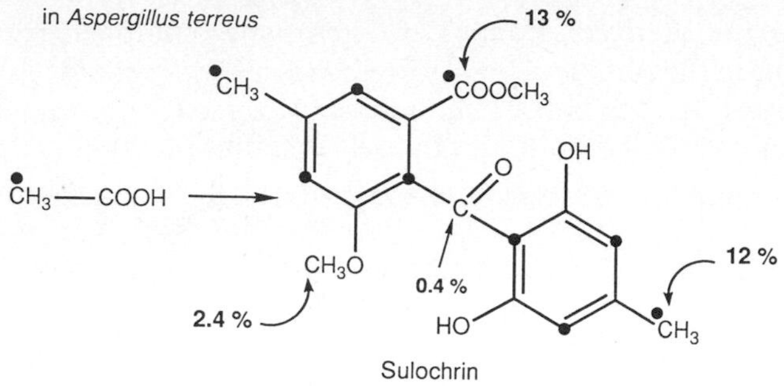

Figure 24 *An example of the methyl group of acetic acid serving as the precursor of an O-methyl group of a natural product*

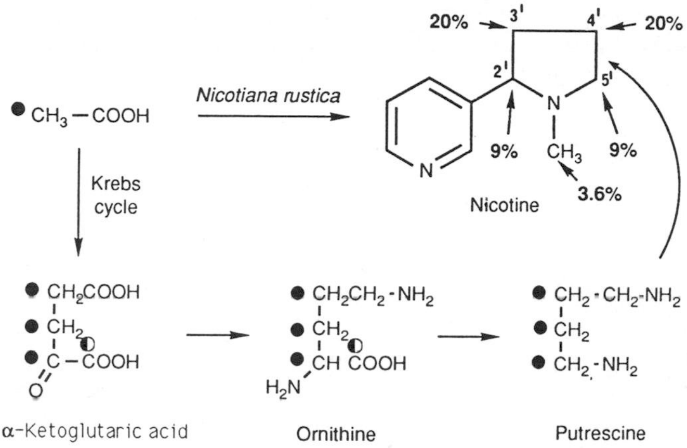

Figure 25 *An example of the methyl group of acetic acid serving as the precursor of an N-methyl group of a natural product*

The mass spectra of the enriched pyrazines from our feeding experiment were also consistent with the ^{13}C NMR data. Figure 26 depicts the mass spectrum of the pyrazine derived from [2-^{13}C]pyruvate. The most prominent molecular ion in the enriched pyrazine was at m/z 155, three mass units more than the unenriched pyrazine. The M-15 at m/z 140 in the enriched pyrazine indicates that the three ^{13}C atoms are in the isopropylpyrazine moiety and none in the O-methyl group.

The mass spectrum of the enriched pyrazine derived from [3-^{13}C]pyruvate (Figure 27) is less clear cut. Considerable unenriched pyrazine (M = 152) is apparently present. However, a significant M + 5 peak at m/z 157 is detected,

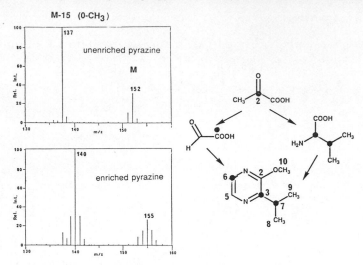

Figure 26 *Mass spectra of unenriched 3-isopropyl-2-methoxypyrazine and that derived from [2-¹³C]pyruvic acid*

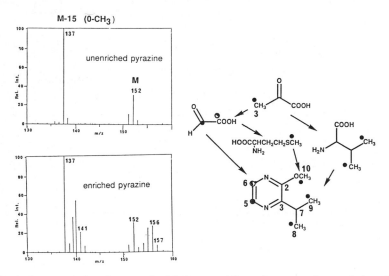

Figure 27 *Mass spectra of unenriched 3-isopropyl-2-methoxypyrazine and that derived from [3-¹³C]pyruvic acid*

indicative of five ¹³C atoms being present in some molecules. Also the peak at *m/z* 141 would result from loss of a ¹³CH₃ group from the methoxy group.

In conclusion, our data are in agreement with those of Gallois, implicating valine in the biosynthesis of 3-isopropyl-2-methoxypyrazine. The origin of C-5 and C-6 is still ambiguous, but our data indicate that this two carbon unit is unsymmetrical. Thus, it cannot be glyoxal as was proposed by Murray.[20]

Acknowledgement.

This work was supported, in part, by a research grant (to E.L.) from the US Public Health Service, National Institutes of Health, GM-13246.

Novel Specific Pathways for Flavour Production

P.S.J. Cheetham

UNILEVER RESEARCH, COLWORTH LABORATORY, COLWORTH
HOUSE, SHARNBROOK, BEDFORD, MK44 1LQ, UK

1 Introduction

The aim of this chapter is to provide some new scientific information about
how flavours can be produced using enzyme-based methods and to place
this information in an industry context, so as to demonstrate the interaction
of the 'market-pull' and 'technical-push' aspects of flavour R&D activities.
In deference to the Societies organizing this conference, it attempts to
emphasize the chemically and botanically related aspects, while trying to
outline some of our current knowledge about pathways producing flavours
and some of the newer biotransformations that are making a scientific and
commercial impact in the industry.

The title prescribed was 'novel specific pathways'; several comments are
necessary. Obviously, from a cost and ease of processing point of view, a
one-step reaction is the ideal, but of course a series of inter-linked reactions
is invariably necessary if considerable modification of the precursor molecules is required in order to form the required flavour molecule(s). Hence
the use of microbial, and in some cases, plant metabolic pathways or parts
of pathways is necessary.

Secondly, the consumer does not really want novel pathways or processes
to be used, but ideally would like nature's traditional pathways to be imitated
as closely as possible in any newly developed flavour production process.[1]
This consumer perception is reflected in the price hierarchy for flavours, *i.e.*
natural > nature-identical > synthetic.

Thirdly, specific pathways producing particular flavour chemicals are
often very useful, but, in addition, non-specific pathways producing a num-

---

[1] P.S.J. Cheetham and S.M.A. Lecchini, in 'Food Technol. Int.', ed. A. Turner, Sterling Publics,
1988, p. 257.

ber of materials with flavour enhancing properties are especially valued because of their complexity and individuality.

However, in trying to develop natural processes, bioscientists have a problem because comparatively few of the metabolic pathways used by microorganisms, fruits, and vegetables, *etc.*, to produce flavours are properly understood, let alone their regulation, control, and the mode of action of the key enzymes. This lack of such basic scientific know-how is quite surprising considering the economic importance of food flavours and also the intrinsic scientific interest of these pathways. Thus we are in the frustrating position of wanting to use plant and microbial pathways in order to produce natural flavours, but often lack the basic scientific information to begin work.

2   Important Characteristics of Flavours

In order to fully appreciate this problem, it is important to emphasize some of the main characteristics of flavours.

(i)   Flavour chemicals are often effective at very low concentrations.
(ii)   The molecules with the most important organoleptic properties are often present at only very low concentrations in the fruit, vegetable, or meat of origin.
(iii)   Flavours, and in particular good quality flavours, are usually complex mixtures of chemicals, many of which contribute to organoleptic value.
(iv)   A very diverse range of different chemicals have flavour properties.
(v)   A wide range of different enzymes are involved in flavour synthesis including not only simple hydrolases and esterases, but also lyases, decarboxylases, deaminases, demethylases, and β-oxidation enzymes, *etc.*
(vi)   Although thousands of molecules have been found to contribute to flavours, only a few hundred are manufactured and used commercially, and of these, only a few are produced on a scale greater than one tonne per annum.
(vii)   Flavours sometimes possess other useful properties such as colour or antimicrobial properties, *etc.*[2]

3   Current Knowledge of Pathways

Some good scientific information has been obtained for some flavour-producing pathways. The pathway from geranyl pyrophosphate to *l*-menthol, which is used specially in dental flavours, has been elucidated (Figure 1). In particular, the metabolism of the key intermediate *l*-menthone into *l*-menthol or *d*-neomenthol and their further metabolism into *l*-menthol acetate and *d*-neomenthol glucoside, and the minor variations in the pathways that

---

[2] P.S.J. Cheetham in 'Biotechnology the Science and the Business' ed. V. Moses and R. Cape, Gordon & Breach, 1991, p. 481.

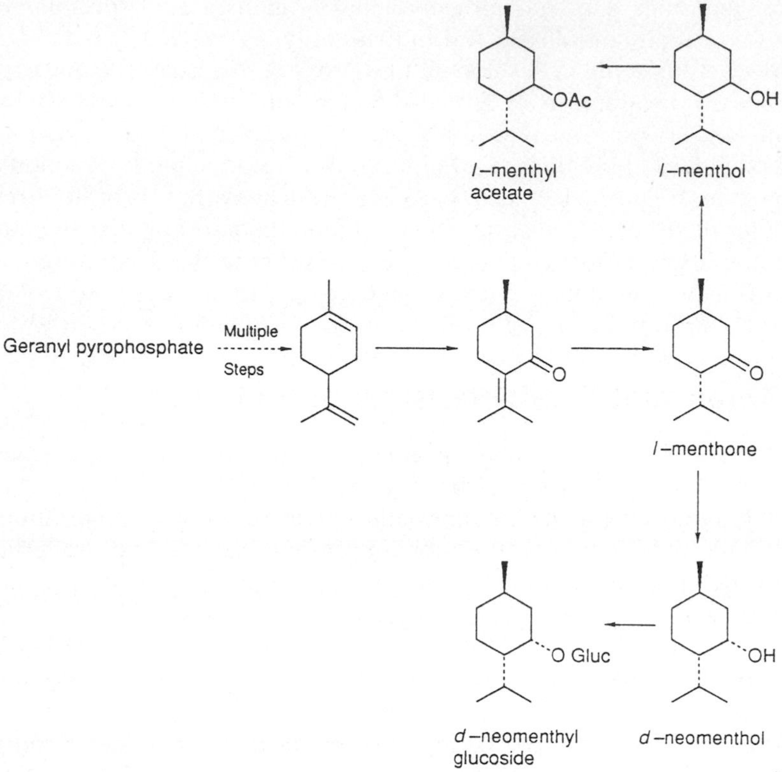

Figure 1 *Final stages in the Synthesis of l-Menthol in Mint Plants*

occur in different mint species have been worked out. Similarly, the formation of 'green' flavour aliphatic alcohols and aldehydes from fatty acids by means of lipoxygenase, hydroperoxide lyases, and alcohol dehydrogenase and isomerase enyzmes has been studied in several species of vegetable and fruit. Thus the different 'green' flavour components present in different plant products can be accounted for by the different substrate specificities of the enzymes present.

Pathways have also been worked out for the formation of onion and garlic flavours, and the dairy flavours diacetyl and the methylketones which are responsible for butter/milk and blue cheese tastes respectively. Diacetyl is formed in micro-organisms such as *Streptococcus diacetylactis* from citric acid, via pyruvic acid, by a number of alternative routes involving condensation of acetyl CoA molecules or decarboxylation of α-acetolactic acid, *etc.* In micro-organisms such as *Penicillium roqueforti*, methylketones are formed from fatty acids by a pathway involving oxidation, hydration, oxidation, and then finally a decarboxylase step. This reaction reduces the concentration of fatty acids, which are toxic to micro-organisms, and so is of obvious advantage to the micro-organism, irrespective of a flavour having been produced (Figure 2).

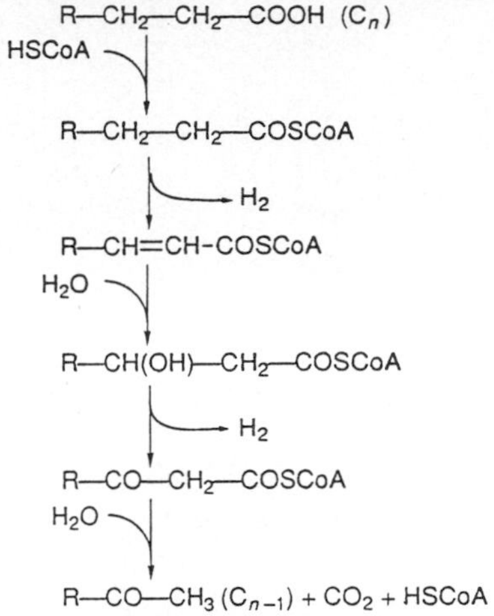

$$R\text{---}CH_2\text{---}CH_2\text{---}COOH \;(C_n)$$

$$R\text{---}CH_2\text{---}CH_2\text{---}COSCoA$$

$$\longrightarrow H_2$$

$$R\text{---}CH{=}CH\text{-}COSCoA$$

$$R\text{---}CH(OH)\text{---}CH_2\text{---}COSCoA$$

$$\longrightarrow H_2$$

$$R\text{---}CO\text{---}CH_2\text{---}COSCoA$$

$$R\text{---}CO\text{---}CH_3\;(C_{n-1}) + CO_2 + HSCoA$$

Figure 2 *Methylketone synthesis by* Penicillium roqueforti

In some cases, this detailed scientific knowledge has been the basis for the development of optimized, large-scale modern processes. For instance, a commercial scale continuous milk souring plant has been developed using *Lactobacillus* and *Streptococcus* strains.[3] Some good work on flavour enzymes and pathways is continuing, for instance, a detailed study of the cyclase enzyme involved in terpene synthesis in plants has been published recently.[4]

4   Recent Examples of Flavour Chemical Synthesis

γ-Decalactone (γ-DL) is a key component of peach and apricot flavours. It can be produced by fermentation of castor oil, in which *ca.* 85 % of the fatty acids are ricinoleic acid. This unsaturated and hydroxylated fatty acid is a particularly good precursor of γ-DL. A number of yeasts such as *Candida*, *Rhodotorula*, and *Sporobolomyces* carry out this fermentation and produce γ-DL in good yield.[5,6,7] The ricinoleic acid is first released by lipase action. It then undergoes four β-oxidation cycles which reduce it from the C_{18} ricinoleic acid to 4-hydroxydecanoic acid (C_{10}). At this stage, further yeast β-oxidation is largely prevented by the proximity of the hydroxyl group to the carboxyl terminus. γ-DL can then be easily formed by lactonization of the hydroxydecanoic acid by heating at acidic pH (Figure 3).

2,5-dimethyl-4-hydroxy-3(2*H*)-furanone (DMHF) is another very important flavour ingredient that is very widely used in fruit and also some

[3] H.L.M. Lelieveld, *Process Biochem.*, 1984, **19** 112.
[4] J.H. Mijazabi and R. Croteau, *Enz. and Microb. Technol.*, 1990, **12**, 841.
[5] M.I. Farbood and B.J. Willis, 1985, US P. 4 560 656.
[6] P.S.J. Cheetham, K.A. Maume, and J.F.M. de Rooij, 1988, Eur. P. Applic. 258 983 A2.
[7] K.A. Maume and P.S.J. Cheetham, *Biocatalysis*, (in press).

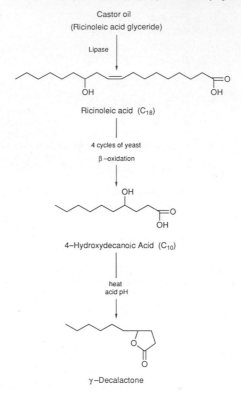

Castor oil
(Ricinoleic acid glyceride)

Lipase

Ricinoleic acid (C_{18})

4 cycles of yeast
β–oxidation

4–Hydroxydecanoic Acid (C_{10})

heat
acid pH

γ–Decalactone

Figure 3   *The fermentation of castor oil into γ-decalactone*

savoury flavours. A scientifically elegant synthesis involving enzyme-mediated carbon–carbon bond formation has been devised using aldolase and triose phosphate isomerase, via a 6-deoxyhexose intermediate.[8] This synthesis is unlikely to be commercially feasible, but a second more practical route involves the generation of rhamnose (6-deoxymannose) by the selective enzymatic or acidic hydrolysis of plant derived flavanoid glycosides such as naringin or rutin (Figure 4).The rhamnose is then heated with an amino acid to form the DMHF flavour. The yield of this final flavour development step is reduced by even small traces of glucose, and since much glucose is also liberated from the flavanoids, a cheap and effective method of selectively removing the glucose has had to be devised.[9, 10] This problem can be solved by three different approaches: by the selective fermentation of the glucose using immobilized *Saccharomyces cerevisiae*, by oxidation to gluconic acid using glucose oxidase, or by fermentation to 5-ketogluconic acid by *Gluconobacter suboxydans*.[9, 10] 5-Ketogluconic acid is important as a precursor for the related monomethylfuranone which possesses good

[8] G.M. Whitesides, D.P. Mazenod, and C.H. Wong, *J. Org. Chem.*, 1983, **48**, 3493.
[9] P.S.J. Cheetham, K.A. Maume, and M.A. Quail, 1987, UK P. Applic. 8 727 222.
[10] P.S.J. Cheetham, S.E. Meakins, and M.A. Quail, 1987 Eur. P. Applic. 8 727 223.

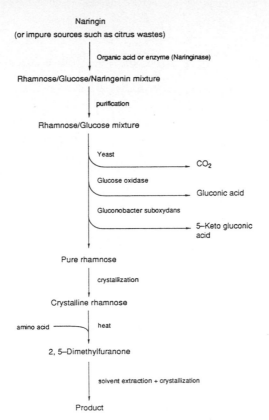

Naringin
(or impure sources such as citrus wastes)

Organic acid or enzyme (Naringinase)

Rhamnose/Glucose/Naringenin mixture

purification

Rhamnose/Glucose mixture

Yeast ——→ CO_2

Glucose oxidase ——→ Gluconic acid

Gluconobacter suboxydans ——→ 5–Keto gluconic acid

Pure rhamnose

crystallization

Crystalline rhamnose

amino acid ——— heat

2, 5–Dimethylfuranone

solvent extraction + crystallization

Product

Figure 4 *Bio-organic procedure for the preparation of 2,5-dimethyl-4-hydroxy-2,3-dihydrofuran-3-one*

savoury flavour properties.[11] It is normally formed by the thermal decomposition of ribose derived from ATP during the cooking of meat. Further reaction with sulphur-containing molecules such as cysteine or methionine gives methylthiophenes which are important components of meat flavours.

Benzaldehyde is well known as having a flavour characteristic of cherries. It can be produced by selective degradation of amygdalin, obtained from sweet almonds, by the action of β-glucosidase and mandelonitrile lyase (Figure 5).

A range of esters can be produced by the oxidative deamination of amino acids such as valine, leucine, and isoleucine in the presence of alcohols. Micro-organisms such as *Geotrichum fragrans* carry out oxidative deamination to produce 2-oxo-isovaleric, 2-oxo-2-methylcaproic acids, which then react with alcohols to form esters[12] (Figure 6).

O-Methylanthranilate is a key flavour character impact molecule of concord/lambrusco grapes. It is also used in perfumes and is an important component

[11] J.F.M. Rooij, 1984, US P. 4 404 409
[12] M.I. Farbood, J.A. Morris, and E.W. Seitz, 1986 Eur. P. Applic. 170 243.

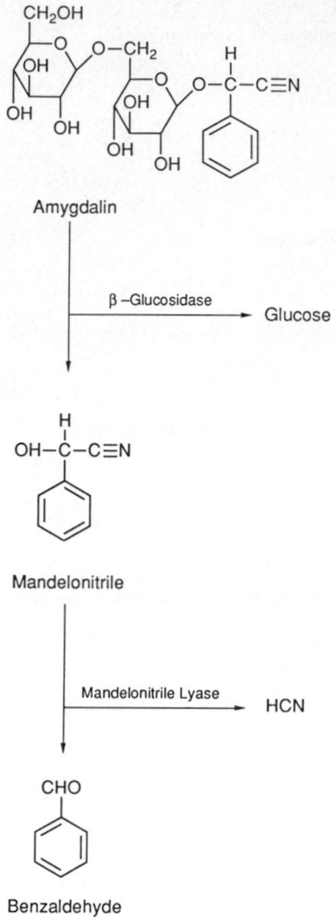
Amygdalin
β-Glucosidase
Glucose
Mandelonitrile
Mandelonitrile Lyase
HCN
CHO
Benzaldehyde

Figure 5 *The enzymic production of natural benzaldehyde*

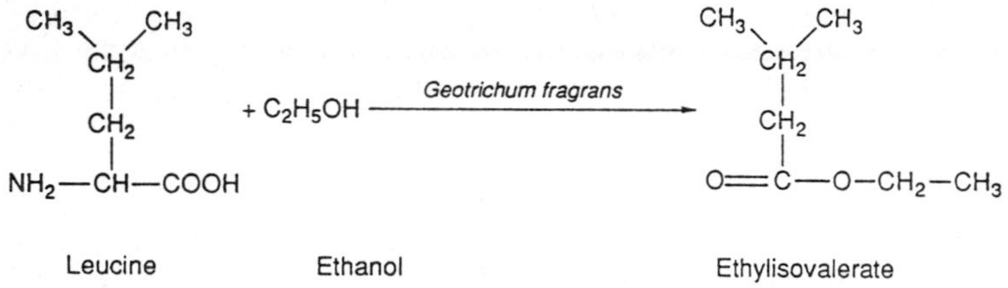
+ C2H5OH
Geotrichum fragrans
Leucine
Ethanol
Ethylisovalerate

Figure 6 *Synthesis of esters by oxidative deamination*

of Poison^{R}. Selective microbial *N*-demethylation of methyl-*N*-methyl-anthranilate, which is readily available from petitgrain mandarin leaf oil; without significant *O*-demethylation, has been reported using *Trametes versicolor* and related micro-organisms[13] (Figure 7).

[13] G.V. Page, B. Scire, and M.I. Farbood, 1989, PCT P. Applic. WO 89/00 203.

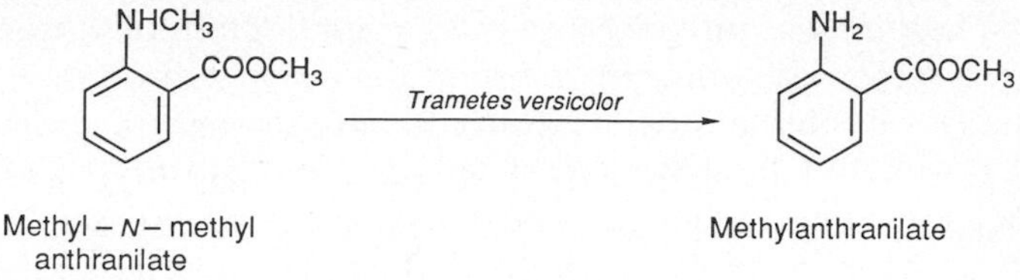

Figure 7 *Natural methylanthranilate by selective* N-*demethylation*

2-Phenylethanol is another material that occurs in both flavours and fragrance materials. It can be produced by the selective degradation of *l*-phenylalanine, via phenylpyruvic acid and phenylacetaldehyde intermediates using deaminase, decarboxylase, and reductase enzymes (Figure 8). Another approach that has been reported recently is *de novo* production by callus cultures of *Rosa damescena*.[14]

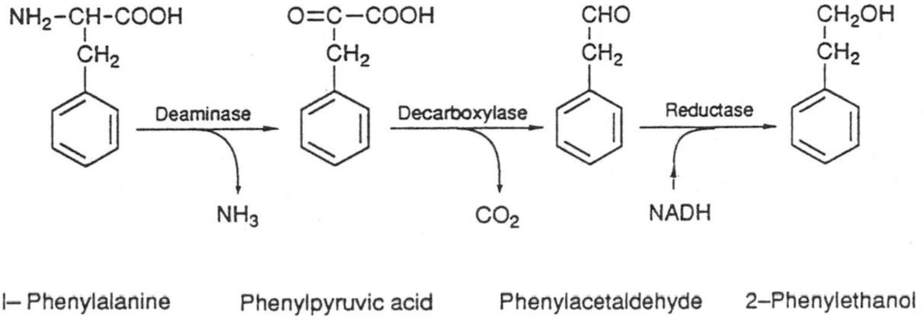

Figure 8 *Catabolic route for the production of 2-phenylethanol in yeast*

The potential of plant cells for flavour production is also being explored. For instance, claims have been made for vanilla flavour production using vanilla callus cells cultured under conditions which select for cells with a high production and secretion capability.[15] Although the pathway whereby vanillin is formed is apparently known, considerable uncertainties still exist about the precise steps involved.[a]

5 Flavour Block Production

In the examples cited so far, the producton of single flavour chemicals has been described. However, several examples of the production of more complex flavour mixtures exist. For instance, accelerated soy sauce processes have been developed in which the raw materials are fermented successively

[14] D.V. Banthorpe, S.A. Branch, I. Poots, and W.D. Fordham, *Phytochem.*, 1988, **27**, 795.
[15] M.E. Knuth and O.P. Sahai, 1989 PCT P. Applic. WO 89/00820.
[a] C. Funk and P.E. Brodelius, *Plant Physiol.*, 1990, **94**, 95.

by three columns of immobilized cells, each containing a mirco-organism known to be important in the traditional soy sauce process.

In the Quest International's savoury flavour process, the biopolymers present in raw materials such as yeast cells are first hydrolysed with proteases, ribonucleases, and carbohydrases. The low molecular weight products of hydrolysis have in many cases good flavour properties, for instance the amino acids and peptides derived from protein and the 5′-GMP nucleotide taste enhancer derived from the yeast's RNA. The flavour quality can then be improved further by additional treatments, such as the use of adenyl deaminase to convert GMP into 5′-IMP, and an anaerobic fermentation step in which sugars are converted to organic acids such as lactic and succinic acids.

6 Elimination of Unwanted Flavours

As well as being useful for the production of desirable flavours biocatalytic processes can also be used to eliminate undesirable tastes. Examples include the debittering of grapefruit juice by the delactonization of limonin and nomilin using the limonin D-ring hydrolase of *Arthrobacter globiformis*[16] (Figure 9), and the debittering of orange juice using naringinase to hydrolyse the glucose and rhamnose sugars from naringin. In dairy products, sulphydryl reductase has been used to reduce the 'cooked' off-flavour of UHT-treated milk.[17] In brewing the diacetyl reductase activity of *Aerobacter aerogenes* has been used to reduce the butter-like off-flavour of some beers.[18]

7 Sweeteners

Sweetness is a very important taste aspect. The high calorific value of normal sugars is an obvious disadvantage, therefore High Intensity Sweeteners (HIS) are of great commercial importance. The enzymatic synthesis of Aspartame[R] from blocked aspartic acid and phenylalanine methyl ester is well known; but recently a technical advance has been made by the isolation of a *Micrococcus caseolyticus* strain that can catalyse the direct synthesis of Aspartame[R] since it reacts selectively with the α-carboxyl group of the aspartic acid and so eliminates the need to protect the β-carboxyl group with a blocking group[19] (Figure 10).

Sucralose[R] (4,1′,6′-trichlorotrideoxy-*galacto*-sucrose) is another HIS which is currently undergoing final trials for regulatory approval. It is about three times as sweet as Aspartame and, most importantly, is heat-stable, so, unlike Aspartame, it can be used in processed foods. Chemical and biochemical

[16] S. Hasegawa and U.A. Pelton, *J. Agric. Food Chem.*, 1983, **31**, 178.
[17] H.E. Swaisgood, *Enz. and Microb. Technol.*, 1980, **2**, 265.
[18] D. Scott, in 'Enzymes in Food Processing' 2nd Edn. ed. G. Reed, Academic Press, New York, 1975, p. 493.
[19] F. Paul., D. Aurial, and P. Monsan, *Enzyme Eng.*, 1989, **9**, 351.

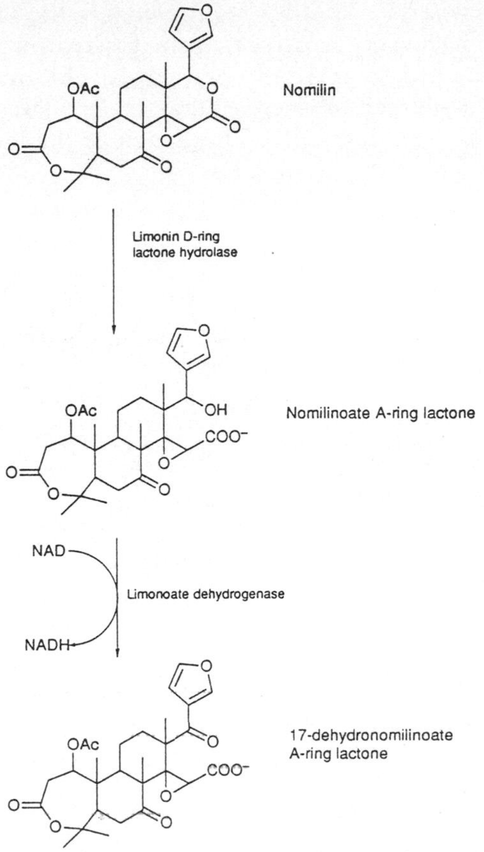

Figure 9 *Debittering of grapefruit nomilin using* Arthobacter globiformis

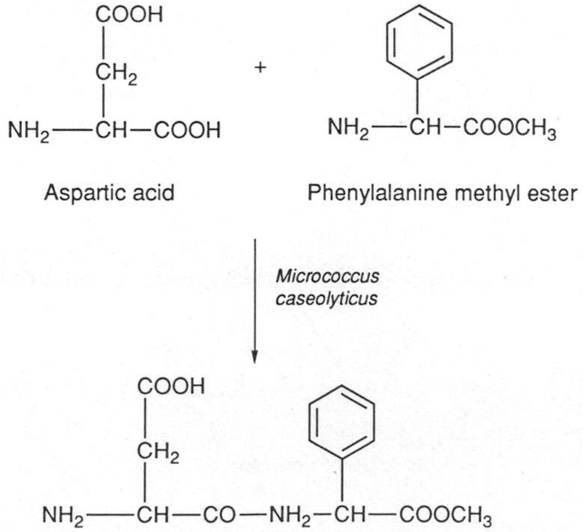

Figure 10 *Enzyme synthesis of Aspartame® using unprotected aspartic acid*

synthesis are possible. The key requirement is the production of intermediates in which the very reactive C_6 primary hydroxyl group of the sucrose molecule is blocked, so as to prevent excessive chlorination and the production of 4,6,1′,6′-tetrachlorosucrose that is less sweet. Several successful syntheses have been developed. These include the fermentation of glucose

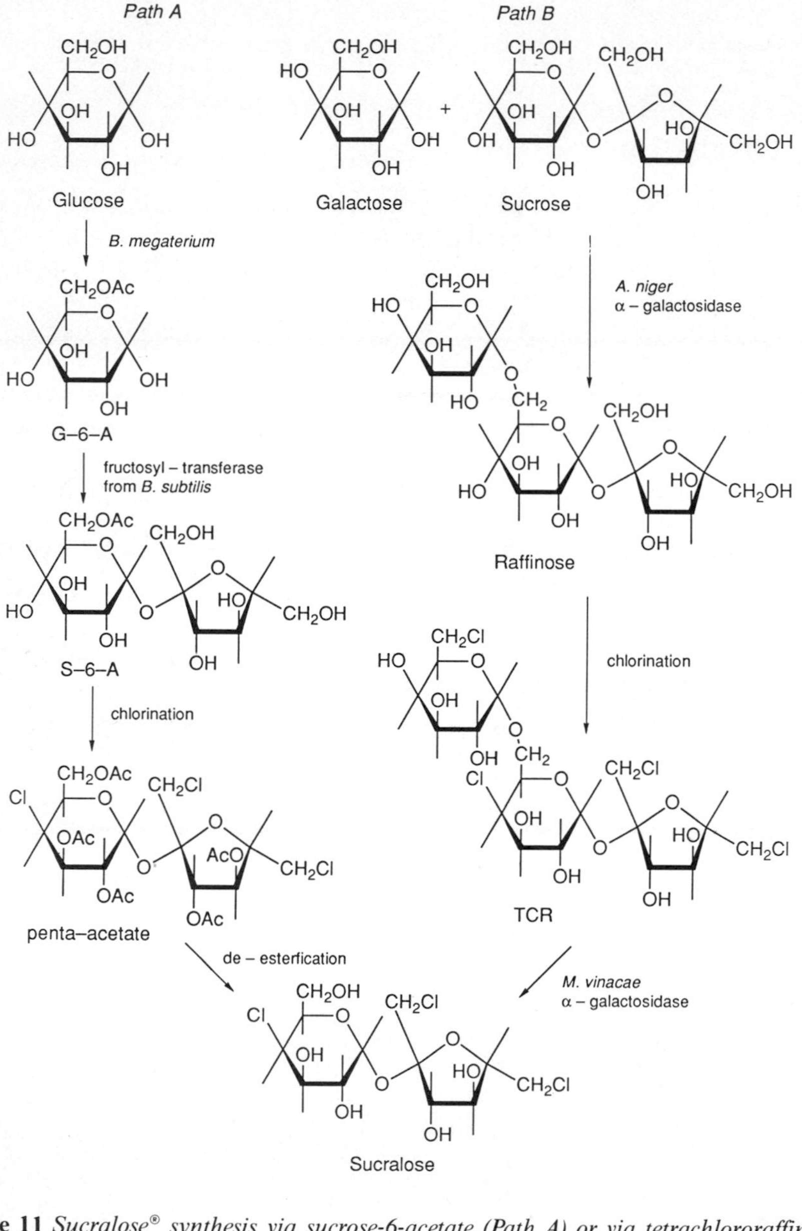

Figure 11 *Sucralose® synthesis via sucrose-6-acetate (Path A) or via tetrachlororaffinose intermediates (Path B)*

by *Bacillus megaterium* to produce glucose-6-acetate, followed by a fructose-transferase reaction to form sucrose-6-acetate, and then chlorination to form sucralose.[20, 21] A second method is by the use of raffinose as the C_6-protected sucrose precursor, followed by chemical derivatization and then selective hydrolysis of the α-1-6 glycosidic bond of the tetrachlororaffinose to generate the sucralose using *Mortierella vinacae* cells[22,23] (Figure 11). Thirdly, sucralose can also be produced via the selective hydrolysis of sucrose-octa-acetate[24] or by the selective acetylation of sucrose by particular lipases.[25]

8   Discussion

In order to build on these successes in the future, new technological developments are required. These include a much improved knowledge of plant and microbial metabolism and its regulation and control, cost-effective genetic engineering techniques, an increased range of industrially suitable biocatalysts, and improved product purification and isolation techniques.

In case the examples already given are not sufficient proof of the great synthetic power of enzymes and micro-organisms, one further example is included. In the shikimate pathway that produces phenylalanine (Figure 12), tyrosine, and tryptophan, and other metabolites in plants and animals, one step is the conversion of 3-deoxy-D-*arabino*-heptulosonate-7-phosphate to dehydroquinate. At first sight, it would appear that several enzymes, or at least a multi-enzyme complex would be required to carry out these reactions. However, the enzyme involved is a simple monomeric enzyme with

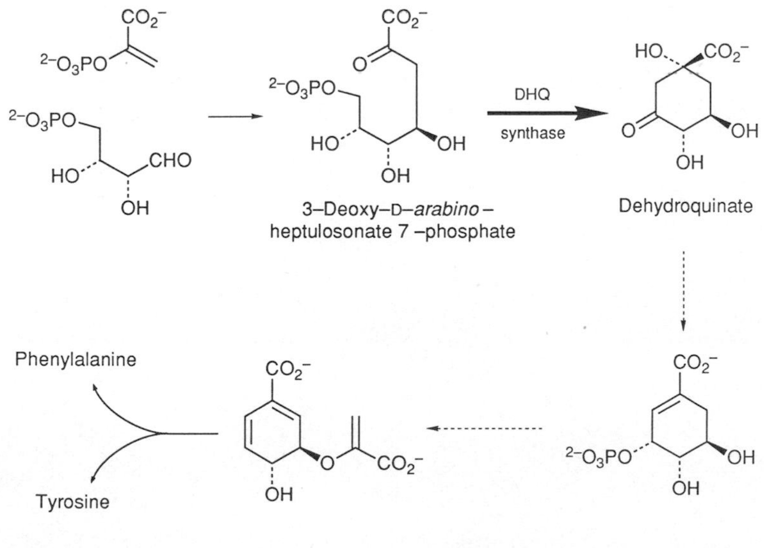

Figure 12 *The shikimic acid pathway illustrating the role of 'DHQ synthase'* (source: J.R. Knowles, Aldrichimica, 1989, **22**, 59)

presumably just one active site. Also, it has an absolute requirement for NAD[+], even though the reaction is redox neutral. In fact, it now appears that a single ordinary NAD[+]-dependent dehydrogenase carries out this reaction, but which Nature has ingeniously organized to involve several kinetically feasible and thermodynamically-favourable, spontaneous steps so that the whole enzyme catalysed reaction can proceed effectively.[26]

[20] E.B. Rathbone, A.J. Hacking, and P.S.J. Cheetham, 1986, US P. 4 617 269.
[21] J.D. Jones, A.J. Hacking, and P.S.J. Cheetham, *Biotech. Bioeng.*, in press.
[22] E.B. Rathbone, K.S. Mufti, R.A. Khan, P.S.J. Cheetham, A.J. Hacking, and J.S. Dordick 1987 UK P. Applic. GB 2 181 734 A.
[23] C. Bennett, J.S. Dordick, P.S.J. Cheetham, and A.J. Hacking, *Biotech. Bioeng.*, in press.
[24] A.J. Hacking and J.S. Dordick, 1990, UK P. Applic. 2 224 504 A.
[25] J.S. Dordick, A.J. Hacking, and R.A. Khan 1990 UK P. Applic. 2 224 773 A.
[26] J.R. Knowles, *Aldrichimica Acta*, 1989, **22**, 59

Chiral Flavour Compounds — Properties and Analysis

R. Tressl, W. Albrecht, and J. Heidlas

TECHNISCHE UNIVERSITÄT BERLIN, INSTITUT FÜR
BIOTECHNOLOGIE, FACHGEBIET CHEMISCH-TECHNISCHE
ANALYSE, SEESTR. 13, 1000 BERLIN 65, GERMANY

1 Introduction

Chiral recognition and the importance of the relationship between stereochemical configuration and biological activity is today well accepted not only for ubiquitous substances such as carbohydrates and amino acids but also in pheromone perception and flavour olfaction. Recently, Ohloff[1] and Pickenhagen[2] have summarized some outstanding examples, *e.g.* carvone, menthol, and nootkatone, to demonstrate the relationship between configuration and sensory properties of chiral flavour compounds. For aliphatic γ- and δ-lactones, however, important constituents of fruits such as strawberries, peach, apricot, and mango there exist only slight differences in the sensory thresholds as quantitatively determined for the optical antipodes of γ-octalactone [(S) : $T(H_2O)$ = 30 p.p.b.; (R) : $T(H_2O)$ = 300 p.p.b.].[3]

Although the knowledge of the molecular mechanisms of olfaction has made great progress with the discovery of the 'odorant binding protein' and the detection of complex messenger reactions the reasons for the discrimination of enantiomers, however, are not yet understood.[1]

Due to the differences in sensory properties and based on the knowledge that enzyme catalysed reactions leading to chiral molecules generally proceed with high enantioselectivity, the determination of the stereochemical configuration of naturally occurring chiral flavour compounds has become

[1] G. Ohloff, 'Riechstoffe und Geruchssinn', Springer-Verlag, Berlin, Heidelberg, 1990.
[2] W. Pickenhagen, in 'Flavour Chemistry — Trends and Developments', ed. R.Teranishi, R.E. Buttery, and F. Shahidi, ACS Symp. Series, No. 388, 1989, p. 151.
[3] J. Heidlas, K.-H. Engel, R.G. Buttery, and R. Tressl, unpublished results.

the aim of intensive research activities. With a few exceptions important flavour components are minor constituents in complex matrices. Therefore, the isolation of a single substance in amounts sufficient for more traditional or NMR spectroscopic characterization is often impossible. This problem has been overcome with the development of gas chromatographic (GC) methods which allow the separation of stereochemical isomers in the nanogram range.

After reviewing the different techniques and recent developments in this analytical field, applications are presented which demonstrate that the gas chromatographic enantioseparation is an important tool (*a*) to prove the authenticity of natural aroma compositions, (*b*) for the investigations of biosynthetic pathways, and (*c*) for the determination of the enantioselectivity in the reduction of prochiral ketones, keto-esters, and diketo compounds catalysed by enzyme preparations.

2 Enantioseparation of Chiral Flavour Compounds by Gas Chromatographic Methods

Gas chromatographic resolution of chiral compounds can be achieved either by formation of diastereomeric derivatives followed by chromatography on achiral stationary phases or by 'direct' separation of enantiomers on optically active stationary phases. Although the scope and limitations of the different methods have been discussed in various critical reports,[4-7] the most important techniques with respect to flavour analysis will be summarized.

2.1 Enantioseparation via Diastereomeric Derivatives

Figure 1 presents the structures of reagents which have been applied for the derivatization of alkan-2-ols, alkan-2-yl esters, hydroxy-acid esters, γ-, and δ-lactones. These reagents are either commercially available or readily synthesized. (*S*)-(+)-α-methoxy-α-trifluoromethylphenylacetic acid chloride [(*S*)-MTPA-C1], originally developed by Mosher *et al.*[8] used for stereochemical analyses by NMR spectroscopy proved to be an excellent reagent for the gas chromatographic separation of diastereomeric derivatives of secondary alcohols, 2- and 3-hydroxy-acid esters.[9] For derivatives of 4-hydroxy-acid ethyl esters, obtained after transesterification of γ-lactones with sodium ethoxide, no quantitative separations were achieved.

[4] V. Schurig, *Angew. Chem.*, 1984, **96**, 733.
[5] V. Schurig, *J. Chromatogr.*, 1988, **441**, 135.
[6] K.-H. Engel in 'Gas Chromatography in Food Analysis and Control', ed. R. Wittkowski, Behrs-Verlag, Hamburg 1991, in press.
[7] V. Schurig and H.P. Novotny, *Angew. Chem.*, 1990, **102**, 969.
[9] J.A. Dale, D.L. Dull, and H.S. Mosher, *J. Org. Chem.*, 1969, **34**, 2543.
[9] R. Tressl and K.-H. Engel, in 'Progress in Flavour Research — Proceedings of the 4th Weurman Symposium', ed. J. Adda, Elsevier, Amsterdam, 1985, p. 441.

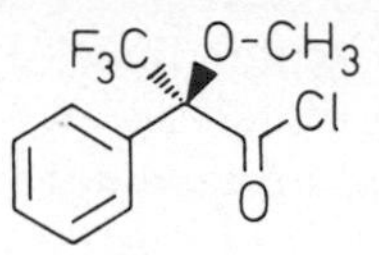

(S)-MTPA-Cl 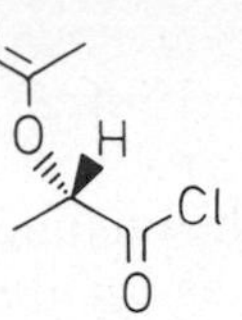   (S)-O-AcL-Cl

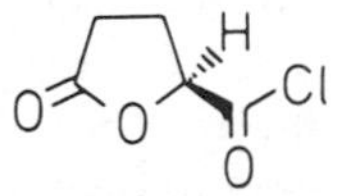

(S)-TOF-Cl 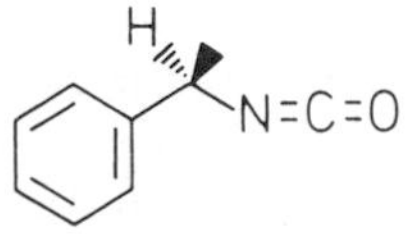   (R)-PEIC

Figure 1   *Reagents for the derivatization of chiral hydroxy components*

In their pioneering work on gas chromatographic enantioseparation, Gil-Av *et al.*[10] formed esters of alkan-2-ols with acetyl lactic acid. Separation was obtained by chromatography on both polar and apolar stationary phases. Mosandl *et al.*[11] revitalized the application of this reagent using the chloride derivatives ((S)-O-AcL-Cl) and separated not only secondary alcohols but also 1,3- and 1,4-diols and 4-hydroxy-acid isopropyl esters, formed from the corresponding 3-hydroxy-acid esters and γ-lactones, respectively.

(S)-(+)-Tetrahydro-5-oxo-2-furancarboxylic acid chloride [(S)-TOF-Cl] has been prepared as a chiral reagent for lactone syntheses.[12] Doolittle *et al.*[13] have tested the quality of this reagent for the gas chromatographic enantioseparation of secondary alcohols. Engel[14] extended the investigations towards the applicability of the simple acid, (S)-TOF, for the separation of 2- and 3-hydroxy-acid esters, and 1,4-diols, formed by reduction of γ-lactones with lithium aluminium hydride (LAH). After transesterification with isopropanol and subsequent reaction with (S)-TOF-Cl Mosandl *et al.* succeeded in the separation of δ-lactones.[15] Slight racemization during the synthesis of this reagent from L-glutamic acid is limiting the routine use of this reagent for the exact determination of enantiomeric ratios of naturally occurring flavour compounds.

With (R)-(+)-1-phenylethylisocyanate [(R)-PEIC], introduced by Pereira *et al.*[16] for the enantioseparation of secondary alcohols, high separation fac-

[10] E. Gil-Av, R. Charles-Sigler, G. Fischer, and D. Nurok, *J. Gas Chromatogr.*, 1966, **4**, 51.
[11] A. Mosandl, M. Gessner, C. Günther, W. Deger, and G. Singer, *J. High Resolut. Chromatogr. Chromatogr. Commun.*, 1987, **10**, 67.
[12] R.E. Doolittle, J.H. Tumlinson, A.T. Proveaux, and R.R. Heath, *J. Chem. Ecol.*, 1980, **6**, 473.
[13] R.E. Doolittle and R.R. Heath, *J. Org. Chem.*, 1984, **49**, 504.
[14] K.-H. Engel, in 'Bioflavour '87', ed P. Schreier, W. de Gruyter, Berlin — New York, 1988, p. 75.
[15] A. Mosandl, C. Günther, M. Gessner, W. Deger, G. Singer, and G. Heusinger, in 'Bioflavour '87' ed. P. Schreier, W. de Gruyter, Berlin — New York, 1988, p. 55.
[16] W. Pereira, V.A. Bacon, H. Patton, B. Halpern, and G.E. Pollock, *Anal. Lett.*, 1970, **3**, 23.

tors were obtained for short chain hydroxy-acid esters.[17] Additionally, aliphatic γ-(C_5–C_{11}) and δ-(C_6–C_{12}) lactones could be separated after reduction with LAH and formation of the corresponding biscarbamates.[18] A decisive improvement for the enantioseparation of γ- and δ-lactones via diastereomeric derivatives has been introduced by Engel *et al.*[19] After amidation of the lactones with butylamine and subsequent reaction with (*R*)-PEIC the homologues γ- and δ-lactones (C_5–C_{12}) can be separated within one derivatization procedure (Figure 2). (*R*)-PEIC is now commercially available in high optical purity so that this method can be applied for the determination of enantiomeric ratios of lactones even in laboratories with standard equipment.

In the described procedures a prerequisite for the formation of diastereomeric derivatives of enantiomers has been the presence of a chemically reactive functional group. Satterwhite and Croteau[20] have overcome this limitation for the enantioseparation of monoterpene hydrocarbons. In two to four step reactions a double bond of the enantiomer was transformed into a keto group which was then derivatized with (2*R*, 3*R*)-2,3-butanediol. The formation of diastereomeric ketals has also been used for the derivatization of δ-lactones[21,22] resulting in a good resolution of the corresponding diastereoisomers, but the applicability of this method is limited by an incomplete reaction.

2.2 Enantioseparation on Optically Active Stationary Phases

The development of the so called 'direct' methods for the gas chromatographic resolution of enantiomers produced three generations of optically active stationary phases with different mechanisms of chiral recognition and different areas of applications.

The thermostable fused silica capillary columns coated with optically active diamide phases (*e.g.* 'Chirasil-Val', 'XE 60-(*S*)-Valine-(*R*)-α-phenylethylamide') can be used for the enantioseparation of various chiral amino compounds, secondary alcohols, 2- and 3-hydroxy-acid esters.[4,23] With a few exceptions derivatization is necessary prior to analysis either to transform non volatile compounds to volatile products or to insert a functional group necessary for the formation of reversible dipole–dipole interactions and hydrogen bondings giving rise to the different retention times of the

[17] R. Tressl, K.-H. Engel, W. Albrecht, and H. Billie-Abdullah, in 'Characterization and Measurement of Flavour Compounds', ed. D.D. Bills and C.J. Mussinan, ACS Symp. Ser. No. 289, 1985, 43.
[18] K.-H. Engel, R.A. Flath, W. Albrecht, and R. Tressl, *J. Chromatogr.*, 1989, **479**, 176.
[19] K.-H. Engel, W. Albrecht, and J. Heidlas, *J. Agric. Food Chem.*, 1990, **38**, 244.
[20] D.M. Satterwhite and R.B. Croteau, *J. Chromatogr.*, 1987, **407**, 243.
[21] G. Saucy, R. Borer, D.P. Trullinger, J.B. Jones, and K.P. Lock, *J. Org. Chem.*, 1977, **42**, 3206.
[22] R. Tressl and K.-H. Engel, in 'Analysis of Volatiles', ed. P. Schreier, W. de Gruyter, Berlin — New York, 1984, p. 323.
[23] W.A. König, 'The Practice of Enantiomer Separation by Capillary Gas Chromatography', Hüthig, Heidelberg, 1987.

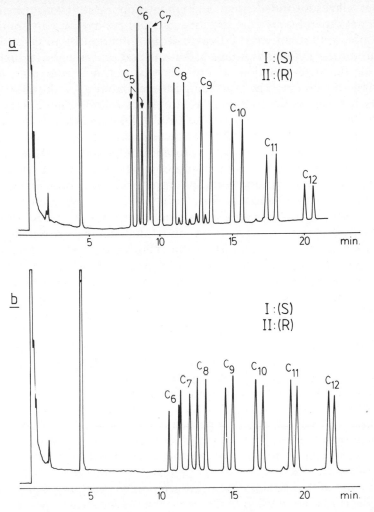

Figure 2 *Capillary gas chromatographic separation of 4- and 5-{(R)-[(1-phenylethyl)carbamoyl]oxy}N-butylcarboxamides derived from γ-lactones (a) and δ-lactones (b). (DB 210, 30m × 0,32 mm i.d., df - 0,25 μm; 220–240 ° C, 5 min isothermal, 1 ° C min⁻¹).*
Reproduced with permission from Reference 19.

enantiomers. With the exception of some chiral alcohols these phases have not been used for investigations of chiral flavour constituents.

In complexation GC which has been introduced by Schurig,[5] chelate complexes of Ni²⁺ or Mn²⁺ and a chiral diketo terpenoid are added to a liquid stationary phase. Reversible interactions between the added chiral selector and the solute, which has to possess suitable chemical functionalities for co-ordination lead to the enantioresolution of several secondary alcohols, ketones, alkyl substituted esters, and ethers. Important flavour constituents, which are separable by complexation GC are methyl 2-methyl butanoate, 1,3-oxathianes and 5-methyl-2-hepten-4-one ('filbertone').[5] Besides high resolution values, a great advantage of this method proved to be its high accuracy for the determination of optical purity even in the case of high enantiomeric excess. The low temperature stability of these stationary phases, however, is limiting their use to highly volatile compounds.

Over the past three years, cyclodextrin (CD) derivatives have been the most widely used stationary phases for the enantioseparation of chiral flavour compounds. The crystalline and hydrophilic α-, β-, and γ-CDs, which have been successfully employed for enantioseparations in thin-layer and high performance liquid chromatography, have to be transformed into hydrophobic derivatives with low melting points and high temperature stability. Recently Schurig and Novotny[7] have summarized the properties of various CD derivatives which have been tested for the gas chromatographic enantioseparations. The coating of the glass or fused silica capillary columns with CDs can be performed in pure form as practised by König and Armstrong and their co-workers. Alternatively, the CD derivatives can be diluted in non-chiral polysiloxanes with appropriate polarity (*e.g.* OV 17, OV 1701) extending the applicable temperature range below the melting point of the CD, introducing the possibility of varying the polarity of the stationary phase. Today a series of capillary columns with CD stationary phases are commercially available and some applications for flavour analysis have been published. With 'Lipodex B' [hexakis(2,5-di-*O*-pentyl-3-*O*-acetyl)-α-CD] Mosandl *et al.*[24] and Bernreuther *et al.*[25] investigated the enantiomeric compositions of γ-lactones isolated from various fruits. Takeoka *et al.*[26] have applied the permethylated β-CD (10 % in OV 1701) for the enantioseparations of naturally occurring terpene hydrocarbons, 2-methylbutanoic acid and the corresponding ethyl ester. On a γ-CD derivative developed by König *et al.*[27] the optical purity of some δ-lactones from coconut and milk products have been determined.[28] Schurig *et al.*[29] reported about the enantioseparation of filbertone, one of the most important constituents in hazelnuts on heptakis(2,6-di-*O*-methyl-3-*O*-trifluoroacetyl)-β-CD (9.5 % in OV 1701).

As a result of intensive and systematic studies on the properties and the separation mechanisms of CD derivatives, Armstrong *et al.*[30] described the resolution of more than 120 pairs of enantiomers, *e.g.* secondary alcohols,

[24] A. Mosandl, U. Hener, U. Hagenauer-Hener, and A. Kustermann, *J. Agric. Food Chem.*, 1990, **38**, 767.

[25] A. Bernreuther, N. Christoph, and P. Schreier, *J. Chromatogr.*, 1989, **481**, 363.

[26] G. Takeoka, R.A. Flath, Th.R. Mon, R.G. Buttery, R. Teranishi, M. Güntert, R. Lautamo, and J. Szejtli, *J. High Resolt. Chromatogr.*, 1989, **12**, 732.

[27] U. Palm, Ch. Askari, U. Hener, E. Jakob, C. Mandler, M. Geßner, A. Mosandl, W.A. König, P. Evers, and R. Krebber, *Z. Lebensm. Unters. Forsch.*, 1991, **192**, 209.

[29] V. Schurig, J. Jauch, D. Schmalzing, M. Jung, W. Bretschneider, R. Hopp, and P. Werkhoff, *Z. Lebensm. Unters. Forsch.*, 1990, **191**, 28.

[30] W.-Y. Li, H.L. Lin, and D.W. Armstrong, *J. Chromatogr.*, 1990, **509**, 303.

vicinal diols, epoxides, γ- and δ-lactones on a 10 m fused silica capillary column coated with octakis (2,6-di-O-pentyl-3-O-trifluoroacetyl)-γ-CD. Despite the great advantages of the CDs it is necessary to draw attention to remarks by these authors concerning column stability. They observed that the column efficiency dropped dramatically when operating at temperatures in excess of 200 ° C. Moisture has to be removed from the carrier gas by an in-line gas purifier because the ester linkage in 3-O-acyl-CD derivatives is susceptible to hydrolysis. Additionally the use of solvents like benzene and toluene which may form inclusion complexes with the stationary phase, should be avoided.

3 Applications

3.1 Analysis of Chiral Flavour Compounds in Fruits

Consumer desire for more 'natural' foods and food products has increased the demand for natural flavours. As this demand exceeds availability, adulteration of foods with nature-identical flavours to increase profitability seems probable. Today two methods for investigating the authenticity of natural food flavours are used. The mass spectrometric carbon isotope ratio measurement ($^{12}C/^{13}C$) can be applied to all flavour compounds. The sample is combusted to CO_2 and the ratio of the ions m/e 45 ($^{13}C^{16}O_2$) and m/e 44 ($^{12}C^{16}O_2$) is measured and expressed in a defined δ-value.[31] For routine use, the naturally occurring values have to be documented for the different plant sources because of variability in isotope discrimination in the CO_2 assimilation by individual plants. This is particularly the case for C_3- and C_4-plants.[32]

The gas chromatographic enantioseparation of chiral flavour compounds can be applied for quality control if enough data of naturally occurring enantiomeric ratios are available. About eight years ago the first results of investigations of chiral constituents from passion fruits, mango, and pineapple demonstrated that generally both enantiomers are biosynthesized in distinct ratios. Exceptions are the optically pure (R)-2-heptyl esters in the red variety of passion fruits.[9] The existence of enantiomeric ratios of chiral constituents led to the question of whether the optical purity is influenced by the state of ripeness and the amount biosynthesized or whether it is a characteristic value for the individual biological system. For an appropriate investigation pineapples were used which contain a series of 3-, 4-, and 5-hydroxy- and acetoxy-acid esters. A five day post-harvest ripening period was accompanied by a reasonable increase in the concentration of flavour

[31] D.A. Krueger, in 'Adulterations of Fruit Juice Beverages', ed. S. Nagy, J.A. Attaway, and M.A. Rhodes, Marcel Dekker, New York — Basel, 1988, 109.
[32] M. O'Leary, *Phytochemistry*, 1981, **20**, 553.

compounds but without significant change of the enantiomeric composition of the chiral constituents.[33]

Such results are particularly important if enantiomeric ratios are to be used for the detection of the adulteration of these components with synthesized 'nature-identical' racemic mixtures.

Aliphatic γ- and δ-lactones are known as important flavour contributing constituents of various fruits.[34] If lactones are biosynthesized in optically active form the gas chromatographic enantioseparation can be used as a very elegant method to prove the authenticity of natural flavours. Table 1 shows some results obtained from investigations of the enantiomeric composition of γ- and δ-lactones, isolated from strawberries, peaches, apricots, nectarines, and from cultures of the micro-organism *Sporobolomyces odorus*. Obviously, the (R)-configuration predominates for γ-lactones with even-numbered chain length, for δ-decalactone and for (Z)-7-decen-5-olide. The constancy of the enantiomeric ratios of the lactones isolated from peaches and apricots has been proved during a six day storage ripening period. Additionally, the optical purity of γ-decalactone in strawberries did not change within a 72 hour storage period although the amount of the lactone increased from 2.4 to 5.2 p.p.m.[35] Based on these results it may be concluded that racemic mixtures of lactones in food based on processed fruits cannot be expected even when many different kinds of fruits are used. The finding of racemic lactones would therefore strongly suggest the presence of 'nature-identical' flavouring.

The routine quality control enantioseparation of lactones can be carried out with modern two-dimensional gas chromatographic equipment. A single compound separated from a complex mixture can be directly transformed on to a second column coated with a suitable CD-stationary phase. In future, the biotechnological production of optically active lactones on a pre-parative scale may well be developed. These products could then be mixed with analogous racemic compounds to adjust to a 'natural ratio'. Several laboratories are therefore investigating the combination of enantioseparation with the mass spectrometric isotope ratio measurement as a means of improving reliability.[36, 37]

3.2   Investigations of Biosynthetic Sequences

Biosynthetic pathways are generally investigated by the application of proposed intermediates to intact organisms, disintegrated cells, or to purified

[33] K.-H. Engel, J. Heidlas, W. Albrecht, and R. Tressl, in 'Flavour Chemistry — Trends and Developments', ed. R. Teranishi, R.G. Buttery, and F. Shahidi, ACS Symp. Series No. 388, 1989, p. 8.

[34] E. Ziegler, 'Die natürlichen und künstlichen Aromen', Hüthig, Heidelberg,1982.

[35] W. Albrecht and R. Tressl, unpublished results.

[36] A. Bernreuther, J. Koziet, P. Brunerie, G. Krammer, N.Christoph, and P. Schreier, *Z. Lebensm. Unters. Forsch.*, 1990, **191**, 299.

[37] A. Mosandl, U. Hener, H.-G. Schmarr, and M. Rautenschlein, *J. High Resolut. Chromatogr.*, 1990, **13**, 528.

Table 1 *Enantiomeric compositions of γ- and δ-lactones isolated from fruits and from cultures of* Sporobolomyces odorus

Compound	Source	Enantiomeric composition[a] (S) (%) (R)		Reference
γ-hexalactone	peach	16	84	(b)
	apricot	16	89	(b)
	nectarine	15	85	(c)
	S.odorus	12	88	(b)
γ-octalactone	peach	12	88	(b)
γ-decalactone	peach	11	89	(b)
	apricot	6	94	(b)
	nectarine	10	90	(c)
	strawberries	< 2	>98	(d)
	S.odorus	< 1	>99	(d)
γ-dodecalactone	peach	< 1	>99	(b)
	apricot	< 1	>99	(b)
δ-decalactone	peach	2	98	(b)
	apricot	5	95	(b)
	nectarine	6	94	(c)
	S.odorus	< 1	>99	(b)
(Z)-7-Decen-5-olide	peach	31	69	(b)
	apricot	5	95	(b)
	S.odorus	27	73	(b)

a   determined according to Reference 18 or Reference 19;
b   W. Albrecht, Ph.D. thesis, Techn.Univ. Berlin, 1991.
c   K.-H. Engel, in 'Bioflavour '87', ed. P. Schreier, W. de Gruyter, Berlin – New York, 1988, p. 75.
d   R. Tressl and W. Albrecht, in 'Biogeneration of Aromas', ed. Th.P. Parliment and R. Croteau, ACS Symp. Series No. 317, 1986, p. 114.

enzymes. Generally, for those experiments the precursors are isotopically labelled. For the investigation of a single enzymatic reaction unlabelled compounds can be used if the conversion is quantitatively measurable.

For the evaluation of oxo-acids as the genuine precursors of chiral pineapple constituents (see above) various 3- and 5-oxo-acids and the corresponding esters were added to submerged pineapple tissue slices.[33] 5-Oxooctanoic acid was transformed into methyl 5-hydroxyoctanoate, methyl 5-acetoxyoctanoate and δ-octalactone in high yields indicating the added substrate to be a precursor. The results of gas chromatographic separation of the diastereomeric biscarbamates formed from δ-octalactone after reduction with LAH revealed that the optical purity of δ-octalactone did not correspond with the biosynthesized compounds. The genuine lactone, isolated from the fruit was almost racemic [(S) : (R) = 47 : 53] in contrast to the high enantiomeric excess of the product obtained after incubation of 5-oxooctanoic acid [(S) : (R) = 8 : 92]. The natural high concentration of the precursor obviously caused participation of carbonyl reducing enzymes in this biotransformation which are not normally involved in the biosynthesis.

Further biosynthetic studies with deuterium labelled substrates revealed that two further pathways leading to δ-octalactone are operative in pineapples (hydration of (Z)-4-octenoic acid; and chain elongation of 3-hydroxy-hexanoic acid). The real activities of these three possible pathways leading to the constant natural enantiomeric ratio of δ-octalactone are still unknown.

For more detailed investigations of the lactone biosynthesis the carotinoid producing red pigmented yeast *Sporobolomyces odorus* was used. As outlined in Table 1 (R)-γ-decalactone isolated from cultures of this organism, is formed in high optical purity. Detailed investigations of the growth, consumption of the carbon sources, and the accumulation of metabolic products revealed that γ-decalactone is a byproduct of fatty acid degradation. With the biotransformation of $[2,2-^2H_2]$-(E)-3-decenoic acid into γ-decalactone, catalysed by whole cells of *S. odorus* the probability of the biosynthesis via this intermediate of the β-oxidation of linoleic acid could be demonstrated.[38] Due to the enzymatically catalysed loss of one deuterium in the course of this biotransformation, hydration of the $(E)-\Delta^3$-double bond leading to 4-hydroxydecanoic acid, the immediate precursor of γ-decalactone, could be excluded. Therefore an epoxidation of (E)-3-decenoic or the corresponding CoA-ester has been assumed. The incubation experiment with racemic $[2,2-^2H_2]$-(E)-3,4-epoxydecanoic acid is schematically summarized in Figure 3. Cells of *S. odorus* transformed the applied double labelled precursor into monodeuterated γ-decalactone, as determined by mass spectrometry. From the gas chromatographic enantioseparation of the derivatized product a non-enantioselective biotransformation could be demonstrated. Additionally, the mass spectrometric investigation of the derivatives revealed that the (S)-γ-decalactone possessed one deuterium whereas the (R)-enantiomer was a mixture of the biosynthesized, unlabelled compound and of the monodeuterated biotransformation product. Based on these results, a biosynthetic pathway (outlined in Figure 4) has been postulated which includes the enantioselective epoxidation of (E)-3-decenoyl-CoA as the decisive step leading to (R)-γ-decalactone.[39]

3.3 Enantioselectivity of Reductions Catalysed by Enzyme Preparations

Comparable to the enzyme catalysed reductions in pineapple, systems in which two or more enzymes with opposite enantioselectivity compete for a substrate constitute a well known phenomenon, an example of which occurs in the catalysed reductions of carbonyl compounds by baker's yeast. This organism was, therefore, chosen as an easily available source for a detailed characterization of the enzymes involved. The separation of two NADPH-dependent oxidoreductases, in the literature designated as '(S)- and (R)-

---

[38] W. Albrecht, Ph.D. Thesis, Techn. Univ. Berlin, 1991.
[39] W. Albrecht and R. Tressl, Z. *Naturforsch.*, 1990, **45c**, 207.

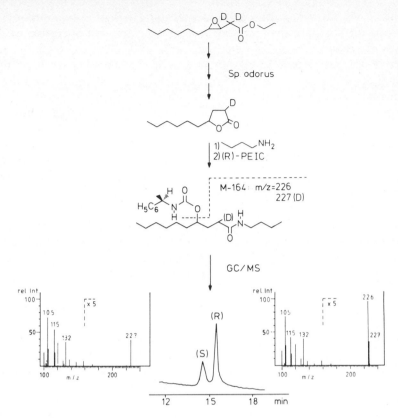

Figure 3 *Biotransformation of [2,2-²H₂]-(E)-3,4-epoxydecanoic acid, catalysed by intact cells of* S. odorus

enzyme', active in the reduction of 3-, 4- and 5-oxo-acids and the corresponding ethyl esters was achieved by means of conventional protein purification methods.[40] The partial separation of both proteins was achieved by ion exchange chromatography on DEAE-cellulose. The degree of separation was monitored by incubation experiments on a preparative scale coupled with spectrophotometrical activity assays. The enzymatically reduced 3-oxo-acid esters were isolated and transformed into diastereomeric MTPA-derivatives for the gas chromatographic enantioseparation.[40]

The '(S)-enzyme' was also active in the reduction of alkan-2-ones, 2,3-butandione (diacetyl) and other diketones.[41] The enantioselectivity of this enzyme for the reduction of diacetyl was confirmed as outlined in Figure 5. The reduction of diacetyl led to one enantiomer as shown by derivatization with (S)-MTPA-Cl. Optically pure reference compounds of acetoin for the determination of the order of elution of the diastereoisomers were not avail-

[40] J. Heidlas, K.-H. Engel, and R. Tressl, *Eur. J. Biochem.*, 1988, **172**, 633.
[41] J. Heidlas and R. Tressl, *Eur. J. Biochem.*, 1990, **188**, 165.

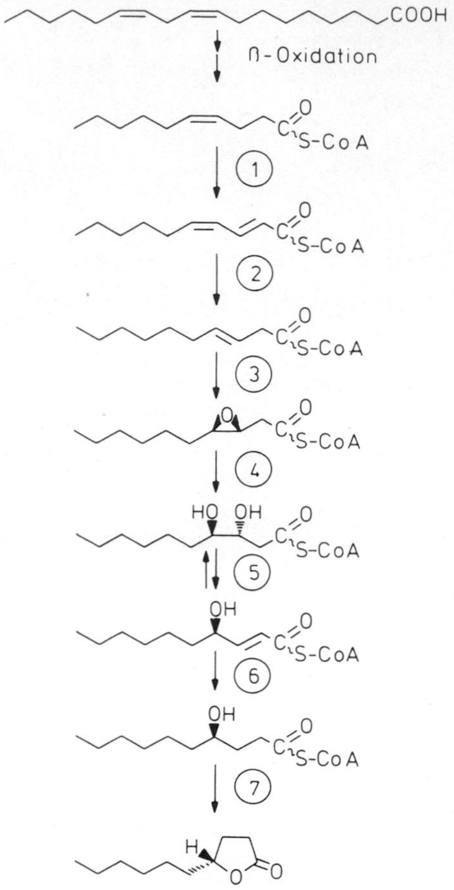

Figure 4 *Proposed bisoynthetic pathway of γ-decalactone in* S. odorus. *1: Acyl-CoA dehydrogenase; 2: 2,4-Dienoyl-CoA reductase; 3: microsomal epoxidase, (not characterized); 4: epoxide hydrolase, (not characterized); 5: Enoyl-CoA hydratase; 6: Enoyl-CoA reductase; 7: lactonization (either enzymatically or thermodynamically catalysed)*

able. The chemical reduction of this enantiomer with sodium borohydride yielded *meso-* and *(S, S)*-2,3-butandiol which could by analysed by capillary GC after derivatization with *(R)*-PEIC (the optical isomers of 2,3-butandiols were available). Therefore the conclusion could be drawn that the investigated isomer was *(S)*.[41]

In contrast to the high optical purity of acetoin, the secondary alcohols obtained after reduction of alkan-2-ones with the '(*S*)-enzyme' were not optically pure. As determined with methylketones of the chain length C_6 to C_{11} the portion of the *(R)*-alcohols varied depending on the substrate.[41] *(S)*- and *(R)*-alkan-2-ols without impurities of the optical antipode were obtained with commercially available preparations of yeast alcohol dehydrogenase

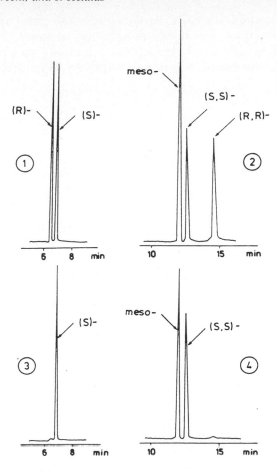

Figure 5 *Gas chromatographic separation of (R)-MTPA esters of acetoin and the (R)-PEIC derivatives of 2,3-butandiol; (1) (R)-MTPA ester of racemic acetoin; (2) (R)-PEIC derivatives of 2,3-butandiol obtained after chemical reduction of racemic acetoin with NaBH₄; (3) (R)-MTPA ester of acetoin formed in a preparative incubation of diacetyl with the purified '(S)-enzyme'; (4) (R)-PEIC derivatives of the 2,3-butandiol obtained after chemical reduction of acetoin formed by the '(S)-enzyme'. (DB210, 30 × 0,32 mm i.d., df = 0,25 μm; (1) and (3): 135 ° C; (2) and (4): 220 ° C.)*
Reproduced with permission from Reference 41.

(YADH) either by reduction of the ketones or by oxidation of the (S)-enantiomer from a racemic mixture of the alcohols.[42] When whole yeast cells are employed the reduction also leads to the (S)-alcohols but only with moderate optical purity.[43] Therefore in addition to the YADH and the '(S)-enzyme' a further oxidoreductase which remains uncharacterized must be active in the formation of (R)-alkan-2-ols.

42 J. Heidlas and K.-H. Engel, and R. Tressl, *Enz. Microbiol. Technol.*, in press.
43 R. MacLeod, H. Prosser, L. Fikentscher, J. Lanyi, and H.S. Mosher, *Biochemistry*, 1964, **3**, 838.

The above examples clearly demonstrate that gas chromatographic enantioseparation is an important tool not only for quality control in foods but also for the isolation and characterization of enzymes necessary for a complete understanding of biosynthetic pathways in natural products.

Application of Natural Product Research in Flavour Creation

J. Knights

PFW (UK) LTD., P.O. BOX 18, 9 WADSWORTH ROAD, PERIVALE, MIDDX. UB6 7JH, UK

1 Introduction

This chapter does not set out to review the methodology used in the analysis of natural food products to elucidate either the mixtures of volatile and non-volatile flavouring substances which are present or the nature of the single substances or complex mixtures generated by biological routes. Rather it outlines the methodology involved in trying to establish the contribution of any individual ingredient, whether a single substance or complex mixture, to the overall perceived flavour of the foodstuff from which it is derived, and the flavour potential which it may have in other types of flavouring.

In recent years the concept of 'naturalness' of food additives including flavourings has stimulated a great deal of research into areas of flavour generation which were previously curiosities. The determination of food marketeers to claim 'natural flavour' for their product as being synonymous with 'good' and 'safe' has resulted in the flavour industry pouring millions of pounds into finding biosynthetic pathways to produce 'natural' versions of many, even most, of the commonly available flavouring substances. It must be obvious that since these substances are chemically identical to their synthetic counterparts, there can be no significant difference in their performance or safety. A major difference is, however, that they are in general many times more expensive than their chemically synthesized versions. As an example, chemically synthesized ethyl butyrate costs about £2.10 kg^{-1} where the comparable prices for biochemically produced material (by the direct esterification of natural ethanol with natural butyric acid) is about £90 kg^{-1}, and for ethyl butyrate isolated physically from orange essence oil the cost is about £2500 kg^{-1}. This does not mean that in every other case the biologically produced material is more expensive and fine chemicals such as ethanol, acetic acid, citric acid, or lactic acid are normally produced commercially by biochemical methods. It is likely that in the long term such labelling claims as 'natural' will be regarded as misleading to the consumer unless the flavouring involved is derived completely from the named foodstuff; moves in this direction are already taking place in the UK[1] and EEC.[2] This does not mean that there is no place for biological methods in the production of flavour ingredients, but rather the opposite. There is the possibility of producing specific flavour ingredients which are either unavailable or uneconomic to produce by normal chemical methods, such as specific optical or structural isomers. Also the production of substances which, if made by chemical methods, are too unstable to allow their isolation before they react further. The present situation with regard to 'naturalness' is totally artificial, whilst the future of biochemical synthesis is very important.

2   Natural Product Investigation

The purpose of undertaking natural food analysis, certainly within the flavour industry, is to try to identify those flavouring substances which contribute towards the characteristic flavour of the food product involved, to determine their contribution and to commercialize their use. The overall scheme which is used is shown in summary in Figure 1. The first three stages, extraction, analysis, and synthesis of the individual substances will not be dwelt upon here but rather, attention devoted to their evaluation and utility. However, for an extraction to be valid, it is essential that the flavour and odour characteristics of the original foodstuff are maintained. Thus the researcher should be working alongside a professional flavourist to ensure that these characteristics are not unduly degraded. At the analysis stage, certainly if it is to be carried out by combined gas chromatography/mass spectrometry (GC/MS), it is necessary that a flavourist, preferably one with experience of smelling the eluate of GC capillary columns, is employed to identify those areas of the chromatograms which are interesting organoleptically. It is better if the flavourist cannot see the chart recorder, since it may prejudice his assessment by making him focus on the emergent peaks when, in our experience, many of the most valuable (unknown) compounds are in such low concentration that they fail to produce a significant chart deflection. The result of such a typical analysis done on cooked pork liver is shown in Table 1. The column headed 'identified by PFW' shows those compounds identified in the sections of the GC trace which the flavourist found to be of interest. Thus the 'previously published in liver' column may well correspond to those parts of the chromatogram in which no interest was expressed. The table also indicates the high level of interest by the flavourist in the sulphur- and nitrogen-containing compounds.

---

[1] MAFF News Release 257/89 Guidance on the use of 'Natural' in Food Labelling, 22 June 1989.
[2] EC Flavourings Directive 88/388/EEC, 22 June 1988.

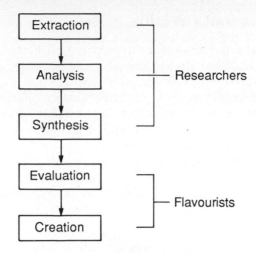

Figure 1 *General scheme for natural product investigation*

Table 1 *Flavour compounds in liver: distribution by chemical class*

Compound Class	Identified by PFW	New to liver	Previously published in liver	Total known in liver	New to food
Hydrocarbons	24	17	29	46	0
Alcohols	22	14	15	29	0
Ethers	11	10	4	14	1
Aldehydes	31	10	29	39	2
Ketones	28	13	26	39	2
Acids	-	-	2	2	0
Esters	35	29	12	41	6
Lactones	-	-	2	2	0
Nitriles	2	2	2	4	0
Phenols	1	0	1	1	0
Furans	10	5	25	30	3
Pyrroles	1	1	5	6	0
Oxazoles	7	7	1	8	5
Thiophenes	11	7	12	19	6
Thiazoles	14	13	3	16	5
Other *S*-Compounds	39	30	15	45	11
Pyridines	15	15	0	15	8
Pyrazines	13	4	35	39	1
Amines	3	3	1	4	1
Total	267	180	219	399	51

Source: PFW unpublished data

3　Evaluation and Creation

Having undertaken the analysis and prepared authentic samples of the compounds which are not already available, the next stage is to evaluate all the substances which have been detected, including those which are well known to flavourists (Figure 2). This should always be undertaken without the identity of the sample being known, otherwise the flavourist's opinion may be prejudiced. For example, every laboratory assistant can recognize benzaldehyde as typical of almond. However, the same compound is present and very important in roast meat flavour; when tasted in a savoury medium it gives very little impression of almond. When a substance is to be evaluated, it should be made up at a suitable concentration in an appropriate simple medium, either acidified sugar syrup or simple broth, and tasted.

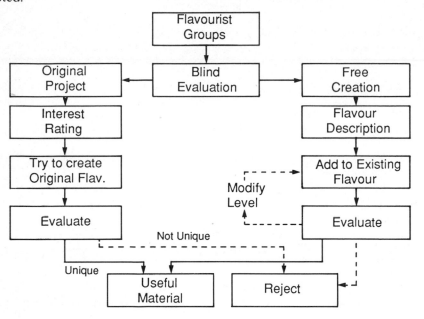

Figure 2　*Evaluation and creation scheme*

The flavourists are then asked a series of questions, the first being in the form of a 'mind map' (Figure 3). They are asked to provide information on the general organoleptic characteristics of the substance and whether these are important and unique. The second series of questions is concerned with its interest rating and suggestions for what other flavours it may be suitable.

A typical computer record for one of the substances identified in cooked liver, 2-acetyl-3(4)methyl pyrrole is shown in Figure 4. As can be seen, the substance has been identified in meat and liver; when tested at 10 p.p.m. in simple broth, it was suggested to be useful in fish and fruit flavours but it achieved only the lowest interest level. The descriptors are given in decreas-

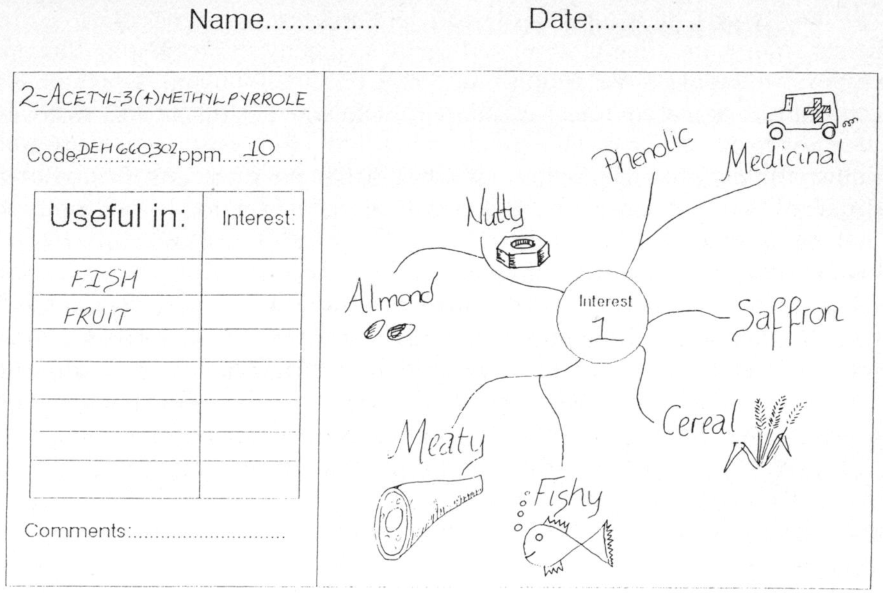

Name..............   Date..............

Figure 3 *Evaluation sheet*

Figure 4 *Flavour creative description*

ing order of perception. The computer program associated with this record allows searches to be conducted on any of the parameters including the descriptors, with choice being available between 1 to 4 or all 8 descriptors. As must be obvious, such a comprehensive record system is extremely useful to the creative flavourist confronted by the problem of finding a substance to provide a particular characteristic note for his original creation.

If the flavour ingredient is of sufficient interest, *i.e.* generally a score greater than 4, an attempt is made to add it directly to an existing flavour which is thought to be appropriate. The modified flavour is evaluated against the original and decisions taken as to whether or not an improvement has been achieved. This process usually takes several trials to decide finally whether the ingredient is useful and that the same effect cannot be achieved using a material which is already in the production inventory. This final step is very important since there is no point on economic grounds of having more stock materials than are really necessary.

The attempt to recreate the flavour of the original project, in this case liver, is much more difficult and in most cases, very close to impossible. Before a start can be made all the identified materials have to be evaluated and then those which appear to be of high interest for liver are put together in approximately the same weight proportion as in the original foodstuff. This level is difficult to determine with any reasonable degree of certainty and analytical artefacts are always present. Generally, this initial flavour compound will have some of the characteristics of liver, but will have many dissimilarities both in concentration and specific flavour notes. For example, the information on the level of phenylacetaldehyde in liver, and indeed in all meat, seems to be almost an order of magnitude higher than can be put into any meat or liver compound without it smelling and tasting perfumed. It is assumed that there must be another substance present which suppresses the effect of phenylacetaldehyde but so far we have been unable to identify it. This is the stage at which the flavourist's knowledge and innate creativity becomes the most important asset in making a commercially acceptable flavouring, by utilizing ingredients which have not been detected in liver but which, from his experience aided by the foregoing computer program contribute to his interpretation of the flavour.

Assuming that one or other of the foregoing evaluation procedures results in the identification of a potentially useful material, it then has to be investigated to determine whether it is possible to make it available as a stock ingredient.

4   Commercialization

In order to make a new flavour ingredient available, three parallel courses of investigation are undertaken by groups representing Research, Development, and Production, and Flavour Creation and Safety Assurance (Figure 5).

The sample which was originally provided for identification and evaluation will have been produced without regard to the practicality or the cost of the method of synthesis, the objective having been to obtain a few milligrams to work on. If the material appears to be useful, a practical synthetic method must be found which is within the cost parameters that can be tolerated by the Flavour Development Group. This obviously requires that the flavour group establishes the maximum level of utilization for the ma-

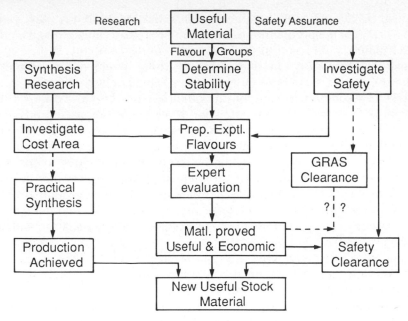

Figure 5 *Commercialization scheme*

terial and thus provides feedback to the R&D group what the maximum practical cost for the material can be. This will eventually determine whether any further synthesis development is required or whether it is unlikely that the cost parameters can be complied with. Assuming that a practical synthesis can be developed no further investigation by R&D and Production is required.

The second parallel course requires that the Flavour Development Group determines whether the chosen material is sufficiently stable in the normal diluants and carrier systems used in production, to make its use a practical proposition. If the material does not have adequate stability, it may be possible to add other ingredients to achieve a sufficient degree of improvement to make it usable. The next stage is for the creative flavourists to prepare experimental flavours and to have them evaluated in normal food media, to establish their character and stability. This procedure of preparation of experimental flavours sounds very simple, but it is very time consuming particularly for flavours which are completely original, that is, where we have no previous experience. Assuming again that the new material is found to be useful, no further work will be undertaken by Flavour Development, pending safety clearance.

The third parallel course is that of Safety Assurance. This is possibly the most difficult area of evaluation as it is largely dependent on the opinion of toxicologists. It requires knowledge of the likely level of use of the material, its normal level of occurrence in foodstuffs, its structure/activity parameters and its likely metabolism. For countries which allow the use of food nature-identical materials without publication, the responsibility for safety becomes that of the producing company, an onorous responsibility which, if miscalculated could result in the bankruptcy of the company. If there is no published information on materials of similar structure, then feeding studies on test animals become mandatory. This is always the case if the material is a complex mixture of microbiological or enzymic origin. If the material is to be made available for use in the US, a submission has to be prepared for the US Flavour Extract Manufacturers Association (FEMA) Expert Panel. This inevitably means that the identity of the ingredient will have to be published, thus divulging the results of our research to our competitors. For many substances we opt not to apply for US clearance in order to keep the identity of the substance confidential. This may be more difficult with the advent of EEC flavour legislation, although the maintenance of confidentiality for a period of at least five years is under discussion with the EC Commission.

It might well be asked what use non-disclosure of a substance's identity could be with modern analytical capability. Although the analysis might indeed be possible, the substances concerned are usually present at extremely low levels and difficult to detect. Further, the decision whether a detected substance has been added deliberately, is present as an impurity, a reaction product, or as an artefact is very difficult. In general, it is much simpler to make that decision if the substance is already published on a positive list of flavouring ingredients. A further possibility is that the substance's use or method of preparation might be patented. Unfortunately, this also means that the identity of the substance has to be divulged and hence does not offer significant protection unless the patentee is prepared to monitor his patent. In the case of single substances, this is not a very successful procedure, but for natural microbiological preparations it is probably essential. Finally, assuming that the material manages to clear all the safety assurance hurdles, together with production and flavour development criteria, a new useful material can be added to the stock for use by the flavourists.

5 Conclusion

Although most of the foregoing has been related to single substances identical to those present in food, a similar procedure is used for substances or preparations of biological or microbiological origin. Biological and microbiological production methods will become increasingly important in the provision of ingredients for the production of flavourings, not because they can be labelled as 'natural' but because they can be made to be more specific and more energy efficient, an important consideration for the future.

Bioreactors for Industrial Production of Flavours: Use of Plant Cells

A.H. Scragg

DEPARTMENT OF SCIENCE, BRISTOL POLYTECHNIC, FRENCHAY, BRISTOL, BS16 1QY, UK

and C. Arias-Castro

DEPARTMENT OF MOLECULAR BIOLOGY, UNIVERSITY OF SHEFFIELD, SHEFFIELD, S10 2TN, UK

1 Introduction

Plant cell cultures have been shown to produce a wide range of flavours characteristic of their plant origin.[1, 2] Flavours such as strawberry, grape, vanilla, tomato, celery, asparagus, spearmint, and quinine[3, 4] are some of the secondary products that have been suggested as possible targets for the industrial use of plant cell cultures. These targets were proposed in the early 1980s when the industrial potential of plant cell cultures first became apparent.[4-6] The industrial use of plant cell cultures was possible because of the high yields of some secondary products and the ability to grow plant cells in bioreactors of large volume.[7-11] The introduction of any new technology is

[1] H.A. Collin, in 'Cell Culture and Somatic Cell Genetics of Plants, Vol. 5', ed. F. Constabel and I.K. Vasil, Academic Press, New York, 1988, p. 569.
[2] Th. Mulder-Krieger, R. Verpoorte, A. Baeheim-Svendsen, and J.J.C. Scheffer, *Plant Cell, Tissue and Organ Culture*, 1988, **13**, 85.
[3] O. Sahai and M. Knuth, *Biotec. Prog.*, 1985, **1**, 1.
[4] M.E. Curtin, *Biotechnology*, 1983, **1**, 649.
[5] A.H. Scragg and M.W. Fowler, in 'Cell Culture and Somatic Cell Genetics of Plants, Vol. 2', ed. I.K. Vasil, Academic Press, New York, 1985, p.103.
[6] J. Berlin, *Endeavour*, 1984, **8**, 5.
[7] W. Barz and B.E. Ellis, *Ber. Dt. Bot. Ges.*, 1981, **94**, 1.
[8] A.K. Panda, S. Mishra, V.S. Bisaria, and S.S. Bhojwani, *Enz. Microb. Technol.*, 1989, **11**, 386.
[9] W. Kreis and E. Reinhard, *Planta. Med.*, 1989, **55**, 409.

generally expensive such that the target compounds chosen are of high value and low volume. Despite considerable efforts and enthusiasm only one process has reached commercial production, that of the production of shikonin by cultures of *Lithospermum erythrorhizon* (Figure 1), although the production of berberine and rosmarinic acid are currently under development.[4]

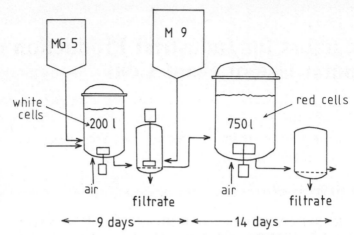

Figure 1 *The two-stage process for the production of shikonin from cultures of* Lithospermum erythrorhizon *developed by the Mitsui Petrochemical Co. MG5 is the growth medium; M9 is shikonin production medium.*
(Reproduced from Reference 4 by permission of Nature Publishing Co.)

The reasons for this slow progress are numerous and are related to the characteristics of plant cell cultures. Plant cells grow slowly and hence high cellular yields are required if overall productivity is to be improved. Although high yields have been obtained for a number of products, the majority of those of commercial interest have proved difficult to produce.[6] Strain selection and screening programmes have worked well with microbial systems but many of the plant products are difficult to assay or to screen. It is, perhaps, not a coincidence that many of the high yielding cell lines so far obtained accumulate products which are either coloured or fluoresce. It has been seen that the yield of secondary product is very variable and is often reduced upon repeated sub-culture.[12, 13] A solution to this variation may be cryopreservation which has been successful for some cultures but has not been widely tested.[14] Secondary products are often produced only after growth has ceased or require an alteration in the culture condi-

[10] G.F. Payne, M.L. Shuler, and P. Brodelius, in 'Large Scale Cell Culture Technology', ed. B.K. Lydersen, Hanser, New York, 1989, p. 193.
[11] A.H. Scragg, in 'Secondary Products from Plant Tissue Culture' ed. B.V. Charlwood and M.J.C. Rhodes, Clarendon Press, Oxford, 1990, p. 243.
[12] P. Morris, K. Rudge, R. Cresswell, and M.W. Fowler, *Plant Cell, Tissue and Organ Culture*, 1989, **17**, 79.
[13] B. Deus-Neumann and M.H. Zenk, *Planta Med.*, 1984, **50**, 423.
[14] L.A. Withers in 'Cryopreservation of Plant Cells and Organs', ed. K.K. Kartha, CRC Press, Florida, p. 243.

tions or medium to induce accumulation.[15, 16] This has led to the development of two-stage processes, like the shikonin process (Figure 1), where the bioreactor runs can be very long.[4] From the considerable work carried out so far it is clear that plant cells grown in culture do not behave in the same way as intact plants and we have a limited understanding of the cellular organization, biochemical changes, and controls that exist.

There are a large number of reports of flavours being produced in cell cultures,[1, 2] but very little information is available concerning the large scale use of plant cells for flavour production. We had hoped to include in this chapter some of our results on the production of glycyrrhizin by *Glycyrrhiza glabra* cultures grown in bioreactors, but, despite promising initial results, the cultures failed to produce glycyrrhizin. This perhaps illustrates the difficulties of such systems but as a consequence we can only review the potential systems available for the large scale production of flavours.

2 The Production of Flavour Compounds by Plant Cell Cultures

The problem of slow growth, low yields and difficulties in the selection and screening of cell lines are the same for flavours as those for pharmaceuticals. However, a number of other problems also occur with flavour components. Flavours are often complex mixtures and are thus far more prone to variation than single pharmaceutical products. Many flavour components are volatile or unstable and are therefore lost rapidly from a culture.[1, 2] Other flavour compounds appear to require some form of cell or organ differentiation for their synthesis and specialized cells for storage.[1,2,17] Therefore, any large scale process will have to be adapted to meet these types of requirement and it is unlikely that a universal process or bioreactor system will become available for flavour production. The processes currently available can be divided into three main types; those using suspension cells,[18–23] those using immobilized cells, and those involving differentiated tissues.[24, 25]

3 Suspension Cultures

In the development of commercial processes utilizing plant cells, much of the work has concentrated on suspension cultures as it was thought that these could be treated like microbial cultures. However, there are differences between plant cell and microbial suspensions which can affect their growth in bioreactors. These are outlined in Table 1 together with the consequences to bioreactor design.

Plant cells, being 40–200 µm long and 10–40 µm in width, are large com-

[15] M.H. Zenk, H. El-Shagi, and M.K. Schulte, *Planta Med. Suppl.*, 1975, 36.
[16] B. Ulbrich, W. Wiesner, and H. Arens, in 'Primary and Secondary Metabolism of Plant Cell Cultures', ed. K-H. Neumann, W. Burz, and E. Reinhard, Springer-Verlag, Berlin, 1985, p. 293.
[17] C. Paupardin, *Congr. Natl. Soc. Sav. Sect. Sci.*, 1976, **101**, 619.

Table 1 *The characteristics of plant and microbial cultures*

Characteristics	Microbial cells	Plant cells	Consequences on large-scale cultivation
Size	2–10 μm	10–200 μm	rapid sedimentation; shear sensitivity
Individual cells	often	not often; aggregates of up to 2mm form	rapid sedimentation; large sample and inoculation ports necessary; formation of micro-environments
Growth rate	rapid	slow; doubling times of 2–5 days	long culture runs; low productivity; maintaining sterility more difficult
Inoculation density	small	large, 10–20 %	large inoculation vessels; reduction in scale-up ratio
Aeration requirements	high	low	low oxygen demand; low K_1a values sufficient for growth
Shear stress sensitivity	insensitive	sensitive/ tolerant	low shear stress bioreactors required; slow impeller speeds

Table taken from Reference 104 by permission of Pergamon Press

pared with micro-organisms and have a rigid cellulose-based cell wall. Both the size and shape of individual cells in a suspension can vary considerably and can change with culture conditions and age.[26–28] Often plant cells go through a phase where the wet weight increases rapidly, the cell expands and the vacuole can take up at least 90 % of the cell. Unlike many micro-organisms, once plant cells have divided they often remain attached forming aggregates which can be up to 2 mm in diameter. The exact origin of these aggregates is unknown as they may form from non-separation but may

[18] Th. Mulder-Kreiger, R. Verpoorte, and A. Baeheim-Svendsen, *Planta Med.*, 1982, **44**, 237.
[19] E.J. Allan and A.H. Scragg, *Biotec. Lett.*, 1986, **8**, 635.
[20] L.A. Anderson, A.T. Keene, and J.D. Phillipson, *Planta Med.*, 1982, **46**, 25.
[21] H.P. Schmander, D. Groger, H. Koblitz, and D. Koblitz, *Plant Cell Rep.*, 1983, **2**, 122.
[22] R. Verpoorte, P.A.A. Harkes, and H.J.G. ten Hoopen, *Acta Leidensia*, 1987, **55**, 29.
[23] E.J. Allen, A.H. Scragg, and K. Pugh, *J. Plant. Physiol.*, 1988, **132**, 176.
[24] H. Becker, *Biochem. Physiol. Pflanzen*, 1970, **161**, 425.
[25] J.K. Webb, D.B. Banthorpe, and D.G. Watson, *Phytochemisty*, 1984, **23**, 903.
[26] H.Tanaka, *Process Biochem.*, 1987, Aug., 106.
[27] A.H. Scragg, P. Bond, F. Leckie, R.C. Cresswell, and M.W. Fowler, in 'Bioreactors and Biotransformations', ed. G.W. Moody and P.B. Baker, Elsevier, London, 1987, p. 12.
[28] A.H. Scragg, P. Bond, and M.W. Fowler, in '6th European Conference on Mixing', Pavia, Italy, AIDIC, 1988, p. 457.

also be formed by aggregation. Here again the nature and size distribution of the aggregates can be affected by cultural conditions and age of culture. The large size of individual cells and the presence of aggregates causes the culture to settle out rapidly.[27,28] This rapid settling will mean that the culture will settle out in areas of poor mixing and this may also block exit and sampling ports. Although the aggregate structure can be loose in nature it may constitute a diffusional barrier, forming its own micro-environments. It has been suggested that the aggregate structure may influence secondary product accumulation due to the development of a different environment.[29] As a consequence the provision of good mixing and absence of dead areas is essential in plant cell bioreactors.

Plant cell suspensions have a very slow growth rate compared with most micro-organisms, with doubling times of 2–6 days. This slow growth means that bioreactor runs will be of 1–3 weeks in length which implies considerable attention to maintaining sterility and will reduce overall productivity considerably. On a production level this will mean that the number of runs that can be obtained from a single bioreactor will be reduced compared with microbial cultures.

Plant cell suspensions have a critical inoculation density below which growth will not occur as the cells appear to require either cell-to-cell contact or unknown substances produced by the cells and exported into the medium. These conditions can be provided by the use of a large inoculum, in general 5–10 % of the total culture volume, which means that inoculation bioreactors will have to be much larger than those used for micro-organisms.

The low volumetric metabolic rates of plant cells result in a low oxygen requirement of 1–30 mM O_2 l^{-1} h^{-1} compared to 5–2000 mM O_2 l^{-1} h^{-1} for micro-organisms.[10] High aeration rates have been shown to effect the growth of plant cells in bioreactors by reducing the levels of carbon dioxide and other volatiles.[30–32] The supply of oxygen to a microbial culture is in general the most critical factor due to the low solubility of oxygen. The supply of oxygen is controlled in stirred-tank bioreactors by both agitation and aeration, with agitation being more important in bubble-breakup and aeration, rather than mixing.

Although early plant cell cultures were grown in stirred-tank bioreactors the difficulties associated with their cultivation were thought to be due to their sensitivity to shear stress.[33,34] Plant cells have been regarded as sensitive to shear stress because of their large size, rigid cell wall, and extensive vacuole. Under the requirement of low aeration and shear sensitivity, the compromise between aeration and agitation is resolved by using low impeller speeds of 50–100 r.p.m. and moderate aeration rates (0.2–0.5 v.v.m.). How-

29 M.H. Shuler, *Ann. N.Y. Acad. Sci.*, 1981, **369**, 65.
30 A. Pareilleux and R. Vinas, *J. Fermentation Technol.*, 1983, **61**, 429.
31 J.P. Duws, G. Feron, and A. Pareilleux, *Appl. Microbiol. Biotech.*, 1988, **25**, 101.
32 P.K. Hegarty, N.J. Smart, A.H. Scragg, and M.W. Fowler, *J. Expl. Bot.*, 1986, **37**, 1911.
33 M. Mandel, *Adv. in Biochem. Eng.*, 1972, **2**, 201.
34 C.C. Dalton, 'Heliosynthase de Aquaculture seminar de Martiques', CNRS, 1978, 1.

ever, in order to increase productivity, plant cell suspensions will have to be cultivated at high densities of 20–30 g dry wt. l^{-1} and this will introduce further problems. It has been suggested that at high densities of up to 40 g dry wt. l^{-1} the culture will be sufficiently viscous to impare mixing.[35] Because of the aggregated nature of plant cell suspensions and the possible presence of extracellular polysaccharides, they have proved to be complicated in their rheological nature and difficult to measure.[36, 37] Wagner and Vogelman[38] showed shear thinning and thixotrophic behaviour for *Catharanthus roseus* and *Morinda citrifolia* cultures. Other cultures have been shown to be non-Newtonian and pseudoplastic.[35] By replacing the normal rotational viscometers with an anchor or turbine impeller, cultures of *C. roseus*, *Helianthus annuus*, and *Acer pseudoplanatus* have been shown to have viscosities of 3–25 mPa at 10–40 g dry wt. l^{-1} [27] (Figure 2A and 2B). Therefore, mixing may not be affected greatly by viscosity.

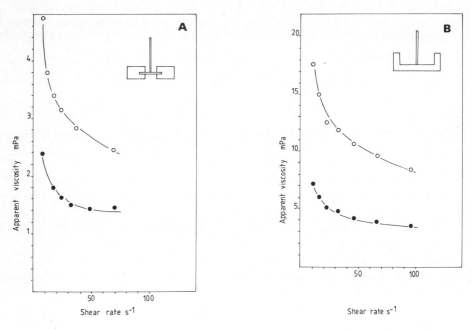

Figure 2 *The apparent viscosity at various shear rates of a culture of* Acer pseudoplanatus *as measured using either;* A. *anchor or* B. *turbine impellers.* A. pseudoplanatus *culture* (●) 200 g l^{-1} wet weight; (○) 400 g l^{-1} wet weight.

[35] H. Tanaka, *Biotec. Bioeng.*, 1981, **23**, 1203.
[36] A.H. Scragg, E.J. Allen, P.A. Bond, and N.J. Smart, in 'Secondary Metabolism in Plant Cell Cultures', ed. P. Morris, A.H. Scragg, A. Stafford, and M.W. Fowler, Cambridge University Press, Cambridge, 1968, p. 178.
[37] A.D. Hale, C.J. Pollock, and S.J. Dalton, *Plant Cell Rep.*, 1987, **6**, 435.
[38] F. Wagner and H. Vogelmann, in 'Plant Tissue Culture and Its Biotechnological Applications', ed. W. Barz, E. Reinhard, and M.H. Zenk, Springer-Verlag, Berlin, 1977, p. 245.

The high rate of settling in plant cell suspensions coupled with their shear sensitivity initiated the study of various methods for agitation and of bioreactor designs. The three basic designs for bioreactor, which are the stirred-tank, bubble column, and airlift bioreactors, are shown in Figure 3. The bubble column and airlift bioreactors use air both to aerate and mix their contents but the inclusion of a draught tube improves mixing greatly. Owing to the lack of impellers, both the bubble column and airlift bioreactors have low shear characteristics.[39, 40] For this reason the airlift bioreactor has been used for the cultivation of plant cells. A number of cultures have been successfully grown in airlift bioreactors up to 200 l in volume.[27, 41]

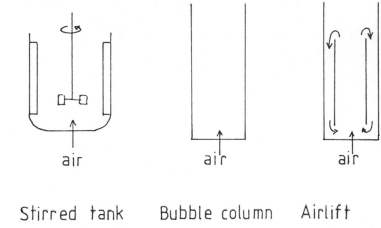

Stirred tank   Bubble column   Airlift

Figure 3   *Basic bioreactor designs*

In addition, a variety of impeller designs have been used in stirred-tank bioreactors in order to achieve good mixing in low shear stress conditions. A comparison of Rushton turbine, anchor, and angled disc turbine impellers in the growth of *Panax ginseng* showed that the angled disc turbine was the most suitable[43] (Figure 4). Ulbrich *et al.*,[16] have shown that a spiral impeller was most effective for the culture of *Coleus* (Figure 4). A cell-lift design was found to be best for cultures of *Glycine max* and *Pinus elliotii*,[43] whereas a large flat-bladed impeller was the most suitable for *Nicotiana tabacum* cultures[44] (Figure 5).

[39] M.H. Zenk, E. El-Shagi, H. Arens, J. Stockigt, E.W. Weiler, and B. Deus, in 'Plant Tissue Culture and its Biotechnological Applications', ed. W. Barz, E. Reinhard, and M.H. Zenk, Springer-Verlag, Berlin, 1977, p. 27.
[40] N.J. Smart and M.W. Fowler, *J. Expl. Bot.*, 1984, **35**, 531.
[41] A.W. Alferman, H. Spieler, and E. Reinhard, in 'Primary and Secondary Metabolism of Plant Cell Cultures', ed. K-H. Neumann, W. Barz, and E. Reinhard, 1985, p. 316.
[42] T. Furuya, T. Yoshikawa, Y. Orihara, and H. Oda, *J. Nat. Prod.*,1984, **47**, 70.
[43] W.J. Treat, C.R. Engler, and E.J. Solter, *Biotec. Bioeng.*, 1989, **34**, 1191.
[44] B.S. Hooker, J.M. Lee, and G. An, *Biotec. Bioeng.*, 1990, **35**, 296.

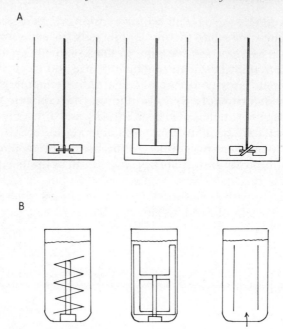

Figure 4   *Various impeller designs. A, turbine, anchor and inclined impellers (as described in Reference 42); B, helical and anchor impeller and airlift (Redrawn from Reference 16)*

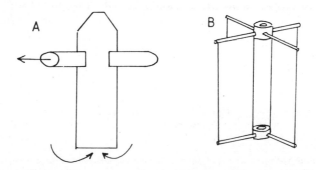

Figure 5   *Alternative impeller designs; A, the cell-lift impeller, (Redrawn from Reference 43) and B, a large flat-bladed impeller (Redrawn from Reference 44).*

All of these studies were based on the assumption that plant cells were sensitive to shear stress. The ability of plant cells to withstand high shear stress has been tested by their exposure to 1000 r.p.m. for up to five hours.[45] Viability was determined by their ability to divide and grow after exposure, as other methods of measuring viability measure only membrane integrity.[46] The results are shown in Table 2. In addition to exposure to short-term shear, long term exposure has also been investigated.[47, 48] Impeller speeds of 500 to 1000 r.p.m. have been used and although at the highest speed growth was somewhat reduced, growth did occur (Figure 6). Inclined impellers have been shown to improve growth at these higher speeds.[47] Therefore, we can now contemplate the use of industrial stirred tank bioreactors for the cultivation of plant cells with the minimum of modifications, where the choice of the right impeller will allow high density cultures to be used.

Table 2 *The effect of shear stress on plant cell cultures*

Cell line	Effect on culture
Short-term exposure	
Catharanthus roseus (3 cell lines)[a]	tolerant
Datura stramonium[a]	sensitive
Helianthus annuus[a]	tolerant
Vitus sp.[a]	tolerant
Solarnum sp.[a]	tolerant
Nicotiana tabacum[a]	sensitive
Picrasma quassioides[a]	tolerant
Long-term exposure	
Catharanthus roseus (IDI)[a]	tolerant
Catharanthus roseus[b]	tolerant
Nicotiana tabacum[b]	tolerant
Tabernaemontana divaricata[b]	tolerant
Cinchona robusta[b]	sensitive

The short term exposure to shear was 5 hours at 1000 r.p.m. and long term was growth at 1000 r.p.m.
[a] Data from Reference 45
[b] Data from Reference 48

[45] A.H. Scragg, E.J. Allen, and F. Leckie, *Enz. Microb. Technol.*, 1988, **10**, 362.
[46] J.M. Widholm, *Stain Technol.*, 1972, **47**, 189.
[47] F. Leckie, A.H. Scragg, and K.C. Cliffe, in 'Progress in Plant Cellular and Molecular Biology', ed. H.J.J. Nijkamp, L.H.W. van der Plas, and J. van Aartijk, Kluwer, Dordrecht, 1990, p. 689.
[48] H.J.G. ten Hoopen, W.M. Gulik, and J.J. Meijer, in 'Progress in Plant Cellular and Molecular Biology', ed. H.J.J. Nijkamp, L.H.W. van der Plas, and J. van Aartijk, Kluwer, Dordrecht, 1990, p. 673.

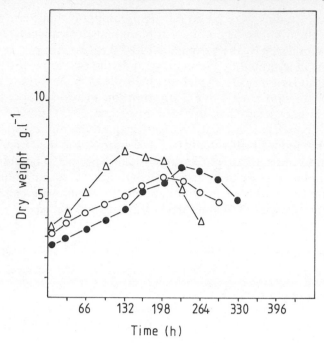

Figure 6   *The increase in dry weight of* Catharanthus roseus *suspension cultivated in a 3 l stirred-tank bioreactor fitted with various types of impeller and run at 1000 r.p.m. (●) turbine; (○) turbine vanes inclined 30 ° to the vertical; (△) turbine vanes inclined 60 ° to the vertical. (Data from Reference 90).*

3.1   Alternative Bioreactor Designs

In order to solve the problems of mixing in low shear stress conditions, alternative bioreactor designs have been used. Tanaka[49] reported the use of a rotating drum bioreactor which has been used to cultivate *Lithospermum erythrorhizon* at volumes of up to 1000 l (Figure 7). The rotating drum gives mixing combined with low shear conditions and the same is true of another bioreactor which uses Taylor–Couette flow to mix the culture[50] (Figure 8). The rotating inner drum creates a series of vortices which mix the culture while aeration is provided by the inner drum which is gas-permeable. This has been used to cultivate *Beta vulgaris* cells. A bioreactor which provides bubble-free aeration has been described by Piehl *et al.*[51] The supply of oxygen is achieved by passing air through a coil made of hydrophobic porous polypropylene. The coil is arranged on a structure which enables the whole coil to be rotated thus mixing the culture. Bioreactors of 1 and 21 l have been developed and used to cultivate *Thalictrum rugosum*.

[49] H. Tanaka, F. Nishijima, M. Suwa, and T. Iwamoto, *Biotec. Bioeng.*, 1983, **25**, 2359.
[50] D.A. Janes, N.H. Thomas, and J.A. Callow, *Biotec. Techniques*, 1987, **1**, 257.
[51] G.W. Piehl, J. Berlin, G. Mollenschott and J. Lehmann, *Appl. Microbiol. Biotec.*, 1988, **29**, 456.

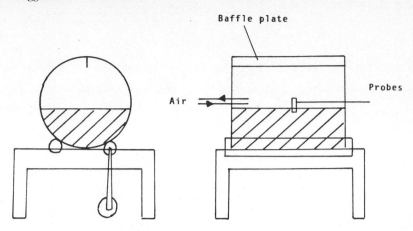

Figure 7 *A schematic diagram of a rotating drum bioreactor.*
(Reproduced from Reference 49, by permission of John Wiley and Sons, Inc)

Figure 8 *A bubble-free annular-vortex membrane bioreactor which exploits Taylor–Couette flow for mixing. (Redrawn from Reference 50)*

3.2 Scale-up

Once a bioreactor design has been shown to function at the laboratory scale, the next stage is to scale-up to production volumes. As scale-up cannot maintain all parameters geometrically, the critical parameters such as mixing will have to be determined. It is also difficult to see some of the more radical designs being scaled-up easily which is clearly not true for stirred-tank bioreactors. Schiel and Berlin[52] have grown *C. roseus* up to a volume of 5000 l (Figure 9) and Westphal[53] has reported the growth of *Echinacea purpurea* up to 75 000 l using stirred-tank bioreactors.

[52] O. Schiel and J. Berlin, *Plant Cell Tissue and Organ Culture*, 1987, **8**, 153.
[53] K. Westphal, in 'Progress in Plant Cellular and Molecular Biology', ed. H.J.J. Nijkamp, L.H.W. van der Plas, and J. van Artijk, Kluwer, Dordrecht, 1990, p. 601.

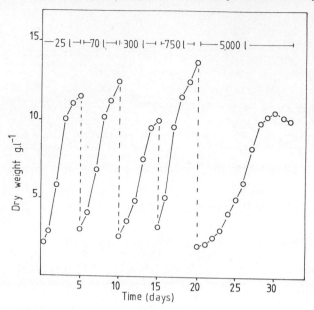

Figure 9   *The scale-up of* Catharanthus roseus *cultures*
(Redrawn from Reference 52 by permission of Kluwer Academic Publishers)

4   Flavour Production by Suspension Cultures

4.1   Quinine

In general flavours are not produced in suspension cultures[17, 24, 25] but there are examples of suspension cultures where flavours may be produced. The first is quinine which is not only used as an anti-malarial drug but in considerable quantities as a bittering agent in soft drinks. Since 1981 a number of laboratories have developed suspension cultures of *Cinchona ledgeriana* and *C. succirubra*, the source of quinine.[18-21] Suspension cultures of *C. ledgeriana* have been grown in 3.5 and 7 l airlift bioreactors[19,21] with doubling times of 5 and 4.8 (0.144 d^{-1}) days respectively. However, the quinine accumulated in the cultures was very low at 0.0867 % and 0.0009 % respectively. Despite these results a design for a large scale fermentation plant has been put forward.[22] The plant was designed to produce 50 kg quinine per year, with a 1 % yield (dry wt.) of quinine, growth rate of 0.164 d^{-1} (dry wt.). The production bioreactor was a 10 000 l airlift, with an inoculation sequence of 100 and 1000 l. Although *C. ledgeriana* has not been tested for shear resistance it may be possible to grow cultures in stirred-tank bioreactors fitted with appropriate impellers. What is obviously needed is to obtain stable high-yielding cell lines of *C. ledgeriana*.

4.2 Quassin

Another bittering agent derived from plants is quassin, a tetracyclic triterpenoid. The normal source is the heartwood of two trees, *Picrasma excelsa* and *Quassia amara*. Suspension cultures have been developed for *Quassia amara* and *Picrasma quassioides*, a related species.[23, 54] *Quassia amara* cultures have been grown in a 7 l airlift bioreactor with a growth rate of 0.133 d^{-1} (t_d = 5.2 days).[54] *Picrasma quassioides* has been tested for tolerance to shear stress and was found to have developed tolerance as its growth rate improved and the culture has subsequently been grown in a stirred-tank bioreactor.[23] Here again the yield of quassin was low (0.01 %) so that the situation is similar to that for *C. ledgeriana* where the process requires the use of a high yielding cell line.

4.3 Glycyrrhizin

The roots of *Glycyrrhiza glabra L.* (Fabaceae), better known as liquorice, contain the natural sweetener glycyrrhizin which is widely used in industry. A suspension culture has been developed using seeds of *G. glabra* and this has been grown in airlift bioreactors up to a volume of 80 l[55] (Figure 10). Growth was somewhat reduced in the 80 l bioreactor. There have been some reports

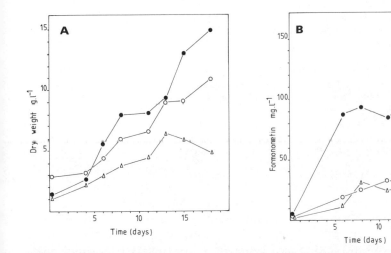

Figure 10 *The growth and formononetin accumulation of* Glycyrrhiza glabra *suspension cultures in airlift bioreactors of 7, 30, and 80 l volume.* (○), *7 l bioreactor, aeration rate 0.14 v.v.m.;* (●), *30 l LHE bioreactor, aeration rate 0.2 v.v.m.;* (△) *80 l LHE bioreactor, aeration rate 0.125 v.v.m. A; dry weight* g l^{-1}, *B; formononetin* mg l^{-1}

[54] A.H. Scragg, S. Ashton, R.D. Steward, and E.J. Allen, *Plant Cell, Tissue and Organ Culture,* 1990, **23**, 165.
[55] C. Arias-Castro, A.H. Scragg, A. Stafford, and M. Rodriguez-Mendiola, *Plant Cell, Tissue and Organ Culture.* 1991, in press.

of the presence of glycyrrhizin in callus and suspension cultures of *G. glabra*,[56] but others have been unable to detect glycyrrhizin.[57] Initial studies indicated that reasonable levels may be found in our cultures. However, on further investigation the compound was shown to be formononetin, an isoflavanoid.

5 Accumulation Sites

Many flavours and essential oils are potentially toxic to the plant in which they are produced such that it is not surprising that they are not detected in suspension cultures. In other cases, these volatile components can be lost or degraded if not removed from the culture. Therefore, the provision of an artificial accumulation site for such volatiles may improve yields. Table 3 lists the additions that have been used to remove compounds from the medium.

Table 3 *The types of artificial accumulation sites used in plant cell suspensions*

Compound	Cell Line	Product	Reference
Miglyol	*Matricaria chamomilla*	essential oils	59
	Thuja occidentalis	monoterpenes	61
RP-8	*Valeriana wallichi*	valepotriates	62
	Pimpinella anisum	anethol	59
XAD-7	*Cinchona ledgeriana*	anthraquinones	69
XAD-4	*Vanilla fragrans*	vanillin	65
charcoal	*Vanilla fragrans*	vanillin	65
hexadecane	*Lithospermum erythrorhizon*	shikonin	66
various solvents	*Lithospermum erythrorhizon*	shikonin	67
sunflower oil	*Capsicum frutescens*	capsaicin	63

Artificial lipophilic compounds, Miglyol 812, a triglyceride with C_8 and C_{10} fatty acids, and RP-8, a modified silica gel with C_8 side chains have been succesfully used to collect essential oils.[58-62] Becker *et al.*,[60] using a two-phase

[56] E. Tamaki, I. Morishita, K. Nishida, K. Kato, and T. Matsumoto, US P. 3 710 512, 1973.
[57] H. Hayashi, H. Fukui, and M. Tabata, *Plant Cell Rep.*, 1988, **7**, 508.
[58] R. Beiderbeck, *Z. Pflanzenphysiol.*, 1982, **108**, 27.
[59] W. Bisson, R. Beiderbeck, and J. Reichling, *Planta Med.*, 1983, **47**, 164.
[60] H. Becker, J. Reichling, W. Bisson, and S. Herold, 'Proceeding of the 3rd Eur. Cong. Biotechnology, Vol. 1', Munich, Verlag Chemie, Weinheim, 1984, p. 1209.
[61] J. Berlin, L. Witte, W. Schubert, and V. Wray, *Phytochemistry*,1984, **23**, 1277.
[62] H. Becker and S. Herold, *Planta Med.*, 1983, **49**, 191.

culture, were able to detect anethole in an anise cell suspension for the first time, although the cell line had been in culture for 18 years. Other compounds used have included charcoal to adsorb sesquiterpenes from *Matricaria chamomilla* cultures[64] and vanillin from *Vanilla fragrans* cultures.[65] Solvents such as hexadecane have been used to remove shikonin from cultures of *L. erythrorhizon*.[66, 67] Ion-exchange resins such as Amberlite XAD-4 and XAD-7 have been used to adsorb vanillin from *V. fragrans* cultures,[65] terpenoids from *T. occidentalis* cultures[68] and anthraquinones from *C. ledgeriana* cultures.[69]

In all cases where the compounds were removed continuously the yield of secondary product increased. Therefore, a bioreactor would either have the adsorbant incorporated into the medium or part of the external system. One such approach is shown in Figure 11 where Amberlite XAD-4 contained in a separate column was used to adsorb excreted terpenoids. Such a system would not be too difficult to incorporate into a bioreactor system but may have the problem of adsorbing some of the growth promotor such that the adsorbant should be chosen carefully.

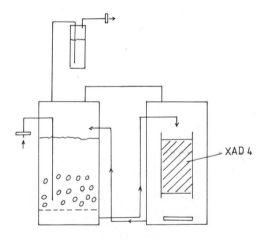

Figure 11 *A system for the continuous extraction of terpenoids from the medium used to culture* Thuja occidentalis. *The adsorbant was Amberlite XAD4 and the bioreactor volume* 1 l
(Redrawn from Reference 68 by permission of Verlag Chemie, Weinheim)

[63] F. Mavituna, A.K. Wilkinson, and P.D. Williams, in 'Separations for Biotechnology', ed. M.S. Verrall and M.J. Hudson, Ellis Horwood, Chichester, 1987, p. 333.
[64] B. Knoop and R. Beiderbeck, *Z. Naturforsch.*, 1983, **38c**, 484.
[65] M.E. Knuth and O.P. Sahai, Patent PCT/US88/02568, 1988.
[66] D.J. Kim and H.N. Chang, *Biotec. Lett.*, 1990, **12**, 289.
[67] H. Deno, C. Suga, T. Morimoto, and Y. Fujita, *Plant Cell Rep.*, 1987, **6**, 197.
[68] E. Forche, W. Schubert, W. Kohl, and G. Höfle, in 'Proceeding of the 3rd Eur. Cong. on Biotechnology, Vol. 1.', Munich, Verlag Chemie, Weinheim, 1984, p. 189.
[69] R.J. Robins, A.J. Parr, and M.J.C. Rhodes, *Biochem. Soc. Trans.*, 1988, **16**, 67.

6 Differentiated Cultures

The second method of improving yield or inducing the formation of flavour is to induce some form of differentiation in the culture. Cultures can be induced to form roots, shoots, or embryos by alterations in the auxin and cytokinin levels. The formation of shoots, or embryos has been shown to induce essential oil accumulation for *Pelargonium tomentosum*, *Pimpinella anisum*, and *Foeniculum vulgare* cultures.[70-72] At present there are no reports of flavour induction with root cultures but such cultures have been shown to produce other types of secondary products.

The growth of root, shoot, or embryo culture in bioreactors imposes a number of restrictions not found with suspension cultures. It is clear that these more organized and larger structures will be more sensitive to shear stress. Akita and Takayama[73] have described a modified stirred-tank bioreactor for the growth of potato tubers which agitated at a low speed (10 r.p.m.) (Figure 12), although the airlift design has also proved to be very successful. The rotating drum bioreactor design which has been used for suspension cultures has also been used for mass embryogenesis, after removal of the baffle (Figure 13). A spin filter bioreactor has also been used for mass

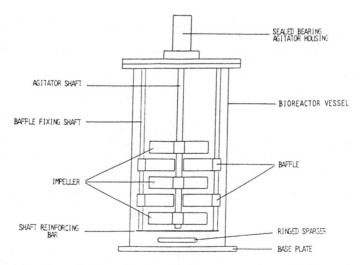

Figure 12 *A bioreactor for the cultivation of plant organ cultures. (Redrawn from Reference 73)*

70 B.V. Charlwood and C. Moustou, in 'Manipulating Secondary Metabolism in Culture', ed. R.J. Robins and M.J.C. Rhodes, Cambridge University Press, Cambridge, 1988, p. 187.
71 D. Ernst, in 'Biotechnology in Agriculture and Forestry, Vol. 7, Medicinal and Aromatic Plants II', ed.Y.P.S. Bajaj, Springer-Verlag, Berlin, 1989, p. 381.
72 G. Hunalt, P. Desmarest, and J. Du Manoir, in 'Biotechnology in Agriculture and Forestry, Vol. 7, Medicinal and Aromatic Plants II', ed. Y.P.S. Bajaj, Springer-Verlag, Berlin, 1989, p. 185.
73 M. Akita and S. Takayama, *Acta Hortic.*, 1988, **230**, 55.

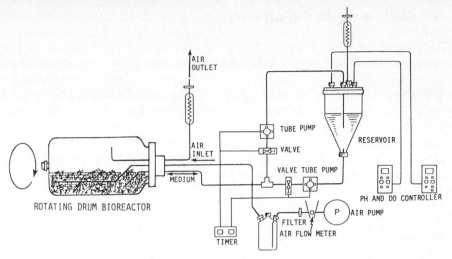

Figure 13 *A rotating drum bioreactor for the growth of shoot cultures. (S. Takayama, personal communication)*

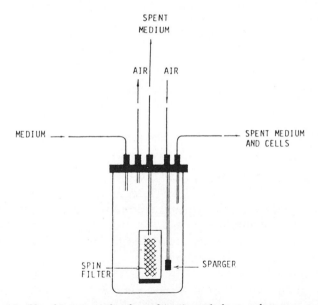

Figure 14 *A spin filter bioreactor for the cultivation of plant embryos.*
(Reproduced from Reference 74 by permission of Plenum Press)

embryogenesis of carrots[74] (Figure 14). The bioreactor consists of a spinning filter mesh which provides agitation and its movement avoids clogging. The system also allows the withdrawal of spent medium without disturbing the culture.

[74] D.J. Styer, in 'Tissue Culture in Forestry and Agriculture', ed. R.R. Henks and K.W. Hughes, Plenum Press, New York, 1985, p. 117.

Plantlets of *Artemisia annua* have been grown in a 2 1 rectangular airlift bioreactor provided with a continuous source of light.[75] The requirement for light has also been addressed in other bioreactors with some designs including a light-introducing draft tube. A very different bioreactor design is the gaseous phase or droplet phase bioreactor, where the culture is grown on a solid support and the medium sprayed from above.[76] The medium spray serves both as a supply of nutrients and aeration. Three designs are shown in Figure 15 where in one case callus material was grown on a conveyor belt system. No extensive study has been carried out on the use of these types of bioreactor for flavour production but simplicity is often an advantage and thus the air driven bioreactors will perhaps be the most successful.

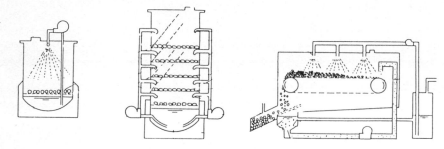

Figure 15 *Three forms of gas phase bioreactors. (Redrawn from Reference 76)*

7   Immobilized Cultures

The term immobilization describes the condition where cells are retained within a matrix or membrane or on an inert support material. Immobilization of plant cells has led to the increased production of secondary products[77] and in some cases their secretion.

Immobilization has a number of advantages over suspension cultures including (*a*) increased cell concentration and hence productivity, (*b*) maintenance of active cells for long periods, (*c*) cell protected from shear, (*d*) medium can be changed rapidly, and (*e*) product or inhibitors removed. The disadvantages are (i) the system can only operate under zero growth conditions unless provision can be made for growth and (ii) the product of interest should be secreted into the medium. If it is not secreted, changes in pH,

[75] J.M. Park, W-S. Hu, and E.J. Staba, *Biotec. Bioeng.*, 1989, **34**, 1209.
[76] K. Ushiyama, H. Oda, and Y. Miyamoto, in 'Abstracts of the VI International Congress of Plant Tissue and Cell Culture', Minnesota University Press, Minneapolis, 1986, p. 252.
[77] K. Lindsey and M.M. Yeoman, *Planta*, 1984, **161**, 495.

ionic strength, addition of permeabilizing agents, or electroporation may be used but this may affect viability.[78-88]

Plant cells can be immobilized in two general ways — either by entrapment or by attachment. Entrapment can be achieved using natural gels (such as agar and alginate), synthetic gels (such as polyacrylamide), trapping behind semipermeable membranes (such as hollow fibre units), and passive entrapment in preformed porous materials (such as polyurethane foam). Covalent attachment and adsorption have also been used but entrapment has been the most popular method due to the sensitive nature of plant cell suspensions.

A considerable range of plant cells have been immobilized mainly using entrapment in gels such as alginate or in preformed polyurethane foams.[78] Those cultures involving flavours are, however, scarce as can be seen in Table 4. In all cases flavour production increased and the product was exported.

Immobilization of plant cells can allow a very varied choice of bioreactor design and process format. The range of these designs can be seen in Table 5 and Figure 16. Only one format has been used so far for flavour production and that is for capsaicin production from *C. frutescens*.[77,83] The cells of *C.*

Table 4 *Flavours produced by immobilized plant cells*

Plant	Product	Type of immobilization	Reference
Capsicum frutescens	capsaicin	polyurethane foam	83
Mentha sp.	terpenes	polyacrylamide	82
Coffea arabica	caffeine	alginate	106

[78] D. Knorr, S.M. Miazga, and R.A. Teutonico, *Food Technol.*, 1985, **39**, 135.
[79] G.F. Payne, M.L. Shuler, and P. Brodelius, in 'Large Scale Cell Culture Technology', ed. B.K. Lydersen, Hanser, New York, 1989, p. 194.
[80] P. Brodelius, in 'Bioreactor Immobilized Enzymes and Cells', ed. M. Moo-Young, Elsevier, London, 1988, p. 167.
[81] K. Lindsey, M.M. Yeoman, G.M. Black, and F. Mavituna, *Febs. Lett.*, 1983, **155**, 143.
[82] D. Galum, D. Arir, A. Dantes, and A. Freeman, *Planta Med.*, 1983, **49**, 9.
[83] F. Mavituna, A.K. Wilkinson, P.D. Williams, and J.M. Park, in 'Bioreactors and Biotransformations', ed. G.W. Moody and P.B. Baker, Elsevier, Amsterdam, 1987, p. 26.
[84] A.K. Wilkinson, P.D. Williams, and F. Mavituna, in 'Plant Cell Biotechnology', ed. M.S. Pais, F. Mavituna, and J.M. Novais, Springer-Verlag, Berlin, 1987, p. 373.
[85] A.J. Lambie, in 'Secondary Products from Plant Tissue Culture', ed. B.V. Charlwood and M.J.C. Rhodes, Clarendon Press, Oxford, 1990, p. 265.
[86] M.J.C. Rhodes, R.J. Robins, J.D. Hamill, A.J. Parr, M.G. Hilton, and N.J. Walton, in 'Secondary Products from Plant Tissue Culture', ed. B.V. Charlwood and M.J.C. Rhodes, Clarendon Press, Oxford, 1990, p. 201.
[87] A. Spencer, J.D. Hamill and M.J.C. Rhodes, *Plant Cell Rep.*, 1990, **8**, 601.
[88] P.D.G. Wilson, M.G. Hilton, R.J. Robins and M.J.C. Rhodes, in, 'Bioreactors and Biotransformations', ed. G.W. Moody and P.B. Baker, Elsevier, London, 1987, p. 38.

Table 5 *Types of bioreactors used for immobilized plant cells*

Bioreactor type	Plant cell line	Reference
Gel entrapped cells		
Stirred-tank	*Digitalis lanata*	91
Airlift	*Digitalis lanata*	91
	Dioscorea deltoides	92
Fluidized-bed	*Catharanthus roseus*	93
	Daucus carota	94
Packed-bed	*Daucus carota*	95
	Catharanthus roseus	96
Polyurethane entrapped cells		
Circulating-bed	*Capsicum frutescens*	83
Fixed-bed	*Capsicum frutescens*	83
	Catharanthus roseus	107
Membrane entrapment		
Tubular or hollow fibre	*Glycine max*	29
Flat-bed	*Glycine max*	103

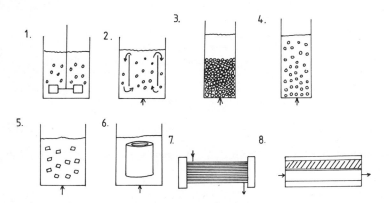

Figure 16 *Various forms of bioreactors used for immobilized plant cells* 1, *stirred-tank*; 2, *airlift*; 3, *packed-bed*; 4, *fluidized-bed*; 5, *circulating-bed for polyurethane foam*; 6, *fixed polyurethane foam*; 7, *hollow-fibre*; 8, *flat-plate membrane.*

frutescens have been immobilized in polyurethane foam either as cubes or as immobilized sheets.[81,83] In the 7 l volume bioreactor using cubes, the cubes of 1 cm³ are restrained while the cells are immobilized and once this has been achieved the cubes are released. Alternatively, the polyurethane foam was used as immobilized sheets and the cell immobilized '*in situ*'. In both cases capsaicin was produced and released into the medium and its formation could be increased by oxygen limitation.[84] In a further study, an extraction column containing sunflower oil was included in this system and the oil acted as an artificial accumulation site thus increasing the overall production of capsaicin.[63]

These types of results have encouraged a study on the feasibility of using such an immobilized plant cell system for the commercial production of capsaicin.[85] The system consisted of a stainless steel reticulate, fixed within the bioreactor of 50 tonnes capacity containing 20 tonne biomass.

At present it is not possible to state which bioreactor design or format is the best for a particular cell line as in practice this is likely to be line dependent, but the possible choice is wide.

8 Transformed Cultures

The transformation of a plant with the plasmid contained within *Agrobacterium rhizogenes* allows the formation of so-called 'hairy-roots' which may exhibit rapid growth and multiple branching. It has been observed that hairy roots are often capable of accumulating high levels of secondary products (see Table 6).[86] At present there are no examples of their use for flavours. There is, however, one report of the use of a transformed culture, that of a shoot culture of *Mentha citrata* transformed by *A. tumefaciens* which was shown to produce terpenes.[87] The bioreactor cultivation of such shoots would have the same constraints as those of non-transformed shoot culture.

Table 6 *Compounds synthesized by transformed root cultures*

Compounds	Genus
Quinoline alkaloids	*Cinchona*
Indole alkaloids	*Catharanthus*
	Clnchona
Tropane alkaloids	*Datura, Hyoscyamus*
	Scopolia, Atropa
Pyridine alkaloids	*Nicotiana*
Betalains	*Beta*
Pyrrolizidine alkaloids	*Securinega*
Steroids	*Solanum*
	Panax
Polyacetylenes	*Tagetes, Bidens*
Naphthoquinones	*Lithospermum*

Reproduced from Reference 86 by permission of Clarendon Press, Oxford.

The hairy root cultures do pose special problems as they grow as tangled masses which if subjected to shear or vigorous mixing will form callus. Growth of hairy roots in bubble columns and airlift bioreactors has shown the nature of this problem with the culture showing negative geotropism and filling the bioreactor with a tangled mass of roots.[88] One solution has been to adopt the gaseous phase or droplet phase bioreactor configuration (Figure 17). This droplet phase bioreactor has been scaled up to 500 l in volume

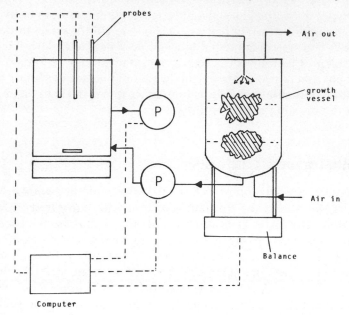

Figure 17 *A droplet phase bioreactor for the growth of hairy root cultures*
(Reproduced from Reference 89, by permission of Kluwer Academic Publishers)

containing a stainless steel matrix for the roots to attach[89] (Figure 18). Another approach has been to provide a mesh of attachment points in an airlift bioreactor so that hairy roots can be evenly distributed throughout the vessel.[104]

[89] P.D.G. Wilson, M.G. Hilton, P.T.H. Meehan, C.R. Waspe, and M.J.C. Rhodes, in 'Progress in Plant Cellular and Molecular Biology', ed. H.J.J. Nijkamp, L.H.W. van der Plas, and J. van Aartijk, Kluwer, Dordrecht, 1990, p. 700.

[90] F. Leckie, A.H. Scragg, and K.C. Cliffe, *Enzyme Microb. Technol.*, 1991, **13**, 296.

[91] P. Markkanen, T. Idman, and V. Kanppinen, *Ann. N.Y. Acad. Sci.*, 1984, **434**, 491.

[92] G.H. Robertson, L.R. Doyle, P. Sheng, A.E. Paulath, and N. Goodman, *Biotec. Bioeng.*, 1989, **34**, 1114.

[93] P. Morris, N.J. Smart, and M.W. Fowler, *Plant Cell. Tissue and Organ Culture*, 1983 **2**, 207.

[94] J.E. Prenosil and H. Pedersen, *Enzyme Microb. Technol.*, 1983, **5**, 323.

[95] I.A. Veliky and A. Jones, *Biotec. Lett.*, 1981, **3**, 551.

[96] P. Brodelius, *Enzyme Eng.*, 1980, **5**, 373.

[97] F. Drawert, R.G. Berger, and R. Godelmann, *Plant Cell Rep.*, 1984, **3**, 37.

[98] H. Itokawa, K. Takeya, and M. Akasu, *Chem. Pharm. Bull. (Tokyo)*, 1976, **24**, 1681.

[99] D. Aviv and E. Galun, *Planta Med.*, 1978, **33**, 70.

[100] T. Suga, T. Hirata, Y. Hirano, and T. Ito, *Chem. Lett.*, 1976, 1245.

[101] T. Furuya, in 'Frontiers of Plant Tissue Culture', ed. T.A. Thorpe, University of Calgary Press, Calgary, 1978, p. 191.

[102] G.J. Lappin, J.D. Stride, and J. Tampion, *Phytochemistry*, 1987, **26**, 995.

[103] M.L. Shuler, J.W. Pyne, and G.A. Hallsby, *J. Am. Oil Chem. Soc.*, 1984, **61**, 1724.

[104] A.H. Scragg, in 'Comprehensive Biotechnology' Vol. 5, ed. G.S. Warren and M.W. Fowler, 1991, in press.

 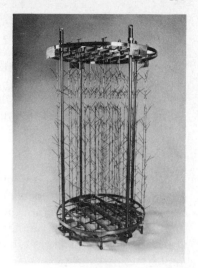

Figure 18 *A* 500 l *pilot plant bioreactor specially designed for the growth of transformed root cultures with a stainless steel matrix to ensure even distribution of inoculum throughout the bioreactor. The design of the matrix enables it to be withdrawn from the biomass after growth*
Photographs by courtesy of Dr. Wilson, Institute of Food Research, Norwich, UK

9   Biotransformation

In all the cases discussed here the process has been developed for the *de novo* synthesis of flavour components but the use of suspended or immobilized cultures for the biotransformation of flavours is an alternative. Examples are shown in Table 7.

Table 7   *Biotransformation of flavour compounds*

Species	Substrate	Product	Reference
Citrus limon	valencene	nootkatone	97
Citrus parodisi	valencene	nootkatone	90
Cannabis sativa	geraniol	citral-a	98
	nerol	citral-b	
Mentha spp.	pulegone	(+)-isomenthone	99
Nicotiana tabacum	linalool	8-hydroxylinalool	100
	dihydrolinalool	8-hydroxydihydrolinalool	
Stevia rebandicina	steviol	steviolbioside	101
		stevioside	
Lavandula	geranial	geraniol	102
angustifolia	neral	nerol	
	cintronellal	citronellol	

Table supplied by Dr. A. Strafford, Plant Science Ltd.

Monoterpenoid biotransformations have been examined in some detail and in particular within cells of species of *Mentha*. There is some evidence of toxicity of monoterpenes but the rate of conversion was rapid and the percentage conversion high. Alfermann *et al.*,[41] have shown with their biotransformation of cardenolides with *Digitalis lanata* cells that bioreactor volumes up to 200 l are possible. Many of the designs described here could be used in a biotransformation system.

10   Conclusions

It is clear that the growth of suspension cultures for the production of flavours can be achieved using a number of bioreactor designs in particular the stirred-tank bioreactor. The growth of roots and shoots either normal or transformed, will require alternative bioreactor designs which may prove difficult to scale-up. However, the real problem is the development of stable high-yielding cell lines.

[105] M.A. Rodriquez-Mendiola, A. Stafford, R. Creswell, and C. Arias-Castro, *Enz. Microb. Technol.*, 1991, **13**, 296.
[106] D. Haldiman and P. Brodelius, *Phytochemistry*, 1986, **26**, 1431.
[107] M.T. Ziyad-Mohamed, Ph.D. Thesis, University of Sheffield, 1991.

Bioreactors for Industrial Production of Flavours: Use of Micro-organisms

J. Crouzet and J.M. Navarro

CENTRE DE GÉNIE ET TECHNOLOGIE ALIMENTAIRES, INSTITUT DES SCIENCES DE L'INGÉNIEUR DE MONTPELLIER, UNIVERSITÉ DE MONTPELLIER II SCIENCES ET TECHNIQUES DU LANGUEDOC, 34095 MONTPELLIER CÉDEX 5, FRANCE

1 Introduction

About 75 % of aroma compounds used as flavorants are natural substances traditionally extracted, in most cases, from plant materials. Technical and economical considerations, as well as progress in biotechnology, have led to the prospect of the use of micro-organisms for the industrial production of aroma. Complex flavours produced by micro-organisms during fermentation processes have been exploited for a long time, and it has been known for more than twenty years that several strains of bacteria, yeasts, or fungi are able to produce, in appropriate medium conditions, several types of aroma compounds such as esters, aromatic compounds, lactones, terpenoids, or pyrazines.[1] The production of these compounds was at first considered a laboratory curiosity, or was used only for taxonomic purposes.[2] Subsequently, numerous works devoted to physiological, biochemical, and analytical studies have been developed,[3-7] and the idea of using micro-organisms for the industrial production of aroma compounds raised.[8]

[1] H.P. Hanssen, in 'Progress in Terpene Chemistry', ed. D. Joulain, Frontières, Gif sur Yvette, 1986, p. 331.
[2] A.F. Halim and R.P. Collins, *Lloydia*, 1972, **34**, 451.
[3] E. Lanza, K.H. Ko, and J.K. Palmer, *J. Agric. Food Chem.*, 1976, **24**, 1247.
[4] E.M.Wilson and V.G. Lilly, *Mycologia*, 1958, **50**, 377.
[5] J.A. Hubball and R.P. Collins, *Mycologia*, 1978, **70**, 117.
[6] H.P. Hanssen and E. Sprecher, in 'Flavour '81', ed P. Schreier, W. de Gruyter, Berlin, 1981, p. 547.
[7] E. Lanza and J.K. Palmer, *Phytochemistry*, 1977, **16**, 1555.
[8] J.A. Maga, *Chem. Senses Flavor*, 1976, **2**, 255.

However, reports related to the study of the production of aroma compounds by micro-organisms using bioreactors are scarce. In the present work, study of the production of monoterpene alcohols, *i.e.* geraniol, nerol, and citronellol by *Ceratocystis moniliformis*, and of decan-4-olide by *Sporobolomyces odorus* is reported.

2   Physiological Considerations

The volatile compounds excreted by micro-organisms are secondary metabolites, and their production is dependent on the composition of the culture medium.[3-6,9,10] For *Ceratocystis moniliformis*,[3] the nature and the quantity of volatile compounds produced, and consequently the odour developed by the culture broth, vary according to the nature of the carbon source (for the same nitrogen source) and *vice versa*. Results obtained in batch culture have shown that, for growth and for production of monoterpene alcohols, the best conditions were obtained when glucose was used as the carbon source and urea as the nitrogen source of nutrients (Table 1).[11] In agreement with the results of Tahara *et al.*,[10] the growth of *Sporobolomyces odorus* and the production of decan-4-olide were maximum in media containing peptone as the nitrogen source.[10] However, the results obtained (Table 2) have shown that, for decan-4-olide production, glucose and fructose are equivalent to mannitol used by these authors.

Table 1   *Effect of carbon and nitrogen source on dry matter, monoterpene alcohol production and sugar consumption by* Ceratocystis moniliformis *in batch culture[a]*

Carbon and nitrogen source	Monoterpene alcohols produced (mg l⁻¹)			Dry matter (g l⁻¹)	Sugar consumed (g l⁻¹)
	Nerol	*Geraniol*	*Citronellol*		
lactose-urea	5.0	1.0	4.0	0.7	15.0
lactose-glycine	1.8	0.1	3.0	0.2	15.0
saccharose-glycine	4.0	16.1	12.0	2.0	–
glucose-urea	30	130	200	5.9	28

[a] N. Jourdain, T. Goli, J.C. Jallageas, C. Crouzet, Ch. Ghommidh, J.M. Navarro, and J. Crouzet in 'Topics in Flavour Research', ed. R.G. Berger, S. Nitz, and P. Schreier, H. Eichhorn, Marzling-Hagenham, 1985, p. 427.

[9] F.M. Young, H.A. Wong, and G. Lim, *Appl. Microb. Biotech.*, 1985, **22**, 146.
[10] S. Tahara, K. Fujiwara, M. Ishizaka, J. Mizutani, and Y. Obata, *Agric. Biol. Chem.*, 1972, **37**, 2585.
[11] N. Jourdain, T. Goli, J.C. Jallageas, C. Crouzet, Ch. Ghommidh, J.M. Navarro, and J. Crouzet, in 'Topics in Flavour Research', ed. R.G. Berger, S. Nitz, and P. Schreier, H. Eichhorn, Marzling-Hagenham, 1985, p. 427.

Table 2 *Effect of carbon and nitrogen source on dry matter. decan-4-olide production and sugar consumption by* Sporobolomyces odorus *in batch culture[a]*

Carbon and nigrogen source	Decan-4-olide produced (mg l^{-1})	Absorbance at 610 nm	Sugar consumed (g l^{-1})
fructose-alanine	1.3	0.8	25.0
saccharose-peptone	2.2	1.1	13.0
glucose-peptone	2.5	1.1	18.0
glucose-asparagine	0.6	0.9	13.9

[a] N. Jourdain, T. Goli, J.C. Jallageas, C. Crouzet, Ch. Chommidh, J.M. Navarro, and J. Crouzet in 'Topics in Flavour Research', ed. R.G. Berger, S. Nitz, and P. Schreier, H. Eichhorn, Marzling-Hagenham, 1985, p. 427.

3 Kinetic Studies

Kinetic studies performed in batch, using sucrose (30 g l^{-1}) and urea (1 g l^{-1}), have shown that the production of monoterpene alcohols by *Ceratocystis moniliformis* was growth-associated (Figure 1). On the other hand, maximum values for yeast growth, as well as for decan-4-olide production, were reached at the end of the linear phase of growth of *Sporobolomyces odorus* in a culture medium containing glucose (30 g l^{-1}) and peptone (0.56 g l^{-1}) at an initial value of pH 6 (Figure 2). When the initial pH was adjusted to 4, the decrease of growth and of decan-4-olide production, and the subsequent increase of pH to about 5.5, was indicative of a modification of the metabolism of the yeast.

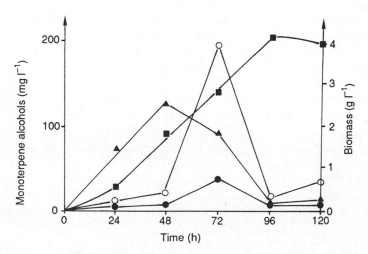

Figure 1 *Biomass (■) and monoterpene alcohol production, [geraniol (△), nerol (●), citronellol (○)] during batch growth of* Ceratocystis moniliformis

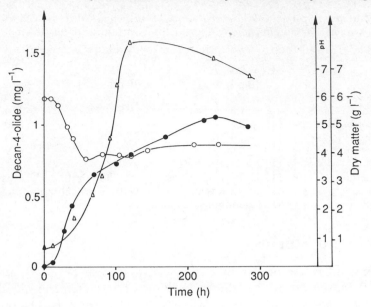

Figure 2   *Biomass* (●), *decan-4-olide production* (△), *and pH variations* (○) *during batch growth of* Sporobolomyces odorus

4   Inhibition by the Products

Another major problem encountered during batch aroma production is inhibition by the products.[12] Different devices have been used in order to overcome this problem.

The repressive effect of terpenoids on *Ceratocystis variospora* metabolism may be eliminated by addition of a lipophilic adsorber, Amberlite XAD 2.[12] According to the author, the yield of terpenoids was increased to about 1.9 g l⁻¹ by circulating the culture medium, contained in a batch fermentor, through an external vessel containing the adsorber.

A semi-continuous fermentation process was used by Kapfer *et al.*[13] for the production of decan-4-olide by *Tyromyces sambuceus*. After four days of culture, removal of medium was carried out by pumping; 50 % of the fermentation broth was pumped out and replaced by the same quantity of fresh medium. Production of the volatile compound was maintained at 220–330 mg l⁻¹ for 70 days.

Another solution, used for the production of related volatile compounds such as ethanol or acetone–butanol, may be explored. Several separation processes, *e.g.* stripping,[14,15] ultrafiltration with hollow fibres or mineral

12 J. Schindler, *I. and E.C. Product. Res. and Dev.*, 1982, **21**, 537.
13 G.F. Kapfer, R.G. Berger, and F. Drawert, *Biotech. Lett.*, 1989, **11**, 561.
14 M.C. Dale, M.R. Okos, and P.C. Wankat, *Biotech. Bioeng.*, 1985, **27**, 943.
15 S.R. Roffler, H.W. Blanch, and C.R. Wilke, *Biotech. Bioeng.*, 1988, **31**, 135.

membranes (Figure 3),[16, 17] coupled ultrafiltration and reverse osmosis,[18] pervaporation,[19, 20] or membrane distillation[21] have been studied during recent years, and their use in aroma production is perfectly possible.

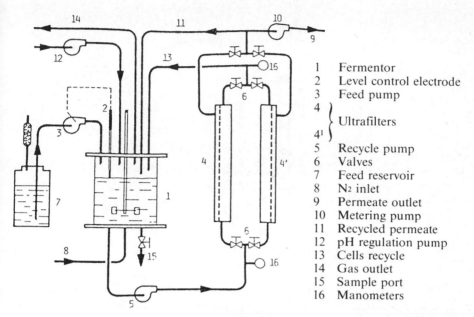

1	Fermentor
2	Level control electrode
3	Feed pump
4	} Ultrafilters
4¹	
5	Recycle pump
6	Valves
7	Feed reservoir
8	N₂ inlet
9	Permeate outlet
10	Metering pump
11	Recycled permeate
12	pH regulation pump
13	Cells recycle
14	Gas outlet
15	Sample port
16	Manometers

Figure 3 *Cell re-cycle apparatus for continuous fermentation coupled with ultrafiltration*

Finally, the use of a continuous reactor is one of the possibilities to be studied. From the results obtained for growth and for the kinetics of production of aroma compounds, three types of reactors, used respectively for the study of continuous production of decan-4-olide and terpenoids, were built and studied.

5 Immobilized Cells Reactor

In the case of decan-4-olide, a considerable biomass at the end of the linear phase is required to attain the highest production level. According to this result, it appeared that immobilization of *Sporobolomyces odorus* was a possible solution.

The yeast was immobilized on an inert support composed of a mixture of wood (75 %) and PVC (25 %) chips. In a first step the continuous reactor,

[16] Y. Nishizawa, Y. Mitani, M. Tamai, and S. Nagai, *J. Ferment. Technol.*, 1983, **61**, 599.
[17] E. Ferras, M. Minier, and G. Goma, *Biotech. Bioeng.*, 1986, **28**, 523.
[18] A. Garcia III, E.L. Iannoti, and J.L. Fisher, *Biotech. Bioeng.*, 1986, **28**, 785.
[19] M. Matsumura and H. Kataoka, *Biotech. Bioeng.*, 1987, **30**., 887.
[20] M.A. Larrayoz and L. Puigjaner, *Biotech. Bioeng.*, 1987, **30**, 692.
[21] H. Udriot, S. Ampuero, I.W. Marison, and U. von Stockar, *Biotech. Lett.*, 1989, **11**, 509.

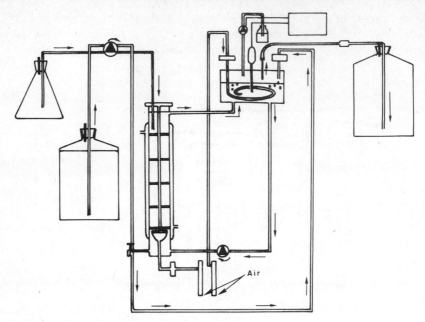

Figure 4   *Immobilized cells reactor, with* pH *regulation, used for continuous production of decan-4-olide by* Sporobolomyces odorus

diagrammatically represented in Figure 4, was used. This reactor consisted of a thermostated EIVS column containing the immobilized yeast and fed by culture broth through a buffer tank used for pH regulation at pH6. The useful volume of the reactor was 1 l.

The reactor was inoculated using 2 l of a three day old pre-culture circulated through the column during five hours. The air flow rate was 0.90 v.v.m. The medium used for feeding contained glucose (10 g l^{-1}) and peptone (2.5 g l^{-1}). The feeding rate was modulated according to the production of decan-4-olide : Q_1, 2.6 ml min^{-1} (dilution rate 0.15 h^{-1}); Q_2, 1 ml min^{-1} (dilution rate 0.06 h^{-1}) and Q_3, 2.2 ml min^{-1} (dilution rate 0.13 h^{-1}). The reactor (Figure 4) was run for a week; the maximum decan-4-olide production was 2 mg l^{-1}, approximately the same as that found in batch culture. The maximum productivity was 0.3 mg l^{-1} h^{-1} (Table 3).

However, the decanolide is produced, not only by the immobilized cells, but also by the free cells present in the buffer tank, and the volume of this tank is more important than the useful volume of the column. From the results obtained in batch culture, showing a pH stabilization at about pH5, a second version of the reactor, without pH regulation, was built (Figure 5). Under these conditions, decan-4-olide is produced only by immobilized cells. The useful volume of this reactor was 430 ml and the air flow rate was 0.7 v.v.m. The same medium was fed at different rates: 15, 30, and 60 ml h^{-1} corresponding to dilution rates of 0.034, 0.069, and 0.138 h^{-1}. As previously, the feed rates were modulated according to the aroma compound production. This reactor was operated for 35 days. Two equilibrium states (Figure

Table 3 *Maximum production and maximum productivity of decan-4-olide by immobilized cells of* Sporobolomyces odorus[a]

Useful reactor volume (l)	Dilution rate (h^{-1})	Air flow rate (v.v.m.)	Maximum production ($mg\,l^{-1}$)	Maximum productivity ($mg\,l^{-1}h^{-1}$)
2.0	0.18	0.5	1.0	0.18
2.5	0.09	0.5	1.75	0.16
2.5	0.06	0.5	2.1	0.14
1.0	0.15	0.9	2.0	0.30

[a] N. Jourdain, T. Goli, J.C. Jallageas, C. Crouzet, Ch. Ghommidh, J.M. Navarro, and J. Crouzet in 'Topics in Flavour Research', ed. R.G. Berger, S. Nitz, and P. Schreier, H. Eichhorn, Marzling-Hagenham, 1985, p. 427.

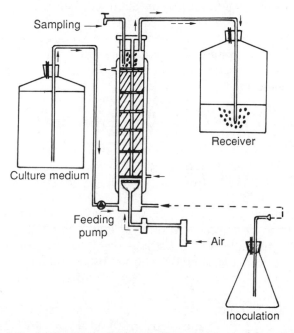

Figure 5 *Immobilized cells reactor, without pH regulation, used for continuous production of decan-4-olide by* Sporobolomyces odorus

6) were reached, corresponding to the production of approximately 0.5 mg l^{-1} and 1.2 mg l^{-1} of decan-4-olide. The decreased production after 650 h of working is clearly associated with contamination, as indicated by the pH increase. The maximum productivity reached is lower than that obtained with the previous version of the reactor. This fact suggests that the production of decan-4-olide by free cells in the buffer tank is more important than the production of this compound by immobilized cells in the column.

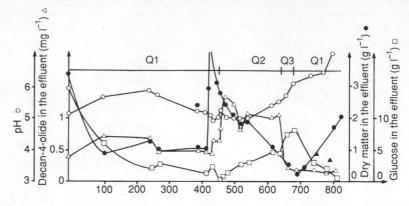

Figure 6   *Time course of continuous production of decan-4-olide by* Sporobolomyces odorus *at various feed rates*: $Q_1 = 15$ ml h^{-1}, $Q_2 = 30$ ml h^{-1}, $Q_3 = 60$ ml h^{-1}

According to the results previously obtained with this kind of reactor,[22] the problems encountered during this study may be related to oxygen transfer. These limitations are generally found in fixed-bed bioreactors in the presence of a gaseous phase, *e.g.* sealing, channelling, high pressure drop, and gaseous retention, with as main consequences, bad mass transfers and reduced useful volumes.

6   Bioreactor with Cross-flow Filtration

In these conditions, ultrafiltration is used for the separation of metabolites and biomass. Preliminary studies were undertaken in order to:
(*a*)   define the effect of hydrostatic pressure on growth and on production of decan-4-olide;
(*b*)   study the mechanical resistance of the yeast to shearing induced by the important mixing necessary to avoid membrane polarization;
(*c*)   study the metabolism of substrates of higher molecular weight than glucose and capable of being retained by the membrane.

Results obtained using pressurized or non-pressurized batch reactors have shown that:
(*a*)   the resistance of the yeast to pressure was less than 2 bars. When the pressure exceeds this value, the growth is slowed down and the secondary metabolism of the yeast appears modified (Figure 7). The necessity to develop low pressure is not a difficulty in so far as a high pressure is not required for the maintenance of a permeative flow rate;
(*b*)   the stress observed under 2 bars is not an oxygen effect. Conversely, in the presence of oxygen at 35 % of saturation, an increase of yeast growth and aroma production was obtained;
(*c*)   dextrin of D.E. 5, 25, and 40 may be used as carbon source by the yeast.

[22]   C. Ghommidh, J.M. Navarro, and G. Durand, *Biotech. Bioeng.*, 1982, **24**, 605.

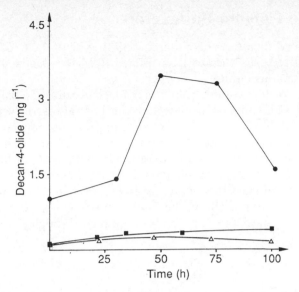

Figure 7 *Effect of pressure on decan-4-olide production in batch culture [atmospheric pressure (△), one bar (●), two bars (■)]*

Based on these results, and preliminary studies using a laboratory ultrafiltration reactor, the reactor shown in Figure 8 was built and tested. The first results obtained with this reactor, with specific regard to its hydrodynamic behaviour, show that the mechanical resistance to shearing strength is good when the circulation rate is less than 1.6 m s^{-1}. A turbulent flow is obtained for a rate of 1 m s^{-1}. It appears that fouling of the membranes by yeasts is not a major problem; the permeate rate decreases from 30 l h^{-1} m^{-2} when the working pressure is established at 15 l h^{-1} m^{-2} after 15 h of work.

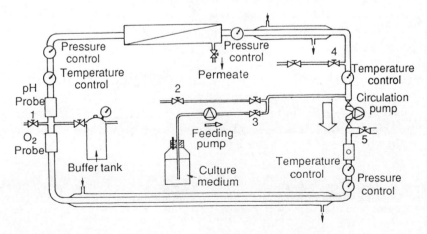

Figure 8 *Cross-flow ultrafiltration reactor studied for the continuous production of decan-4-olide*

7 Bubble Column Bioreactor

Because of the difficulties encountered with the use of the fixed-bed reactor with immobilized cells, the use of a triphasic fluidized reactor (Figure 9) was tried for continuous culture of *Ceratocystis moniliformis*.

This reactor of 1.5 l volume, all glass made, consisted of a column of 0.9 l, a decantor of 0.2 l, and a headspace of 0.4 l.[23] Lateral inserts were fitted with pH and temperature probes connected to appropriate regulators. Fluidization of biomass was achieved using air introduced at the bottom of the reactor. Two different distributors were used; in the first configuration, the distributor was composed of eight syringe needles of 0.8 mm diameter, whereas a sintered glass disk, porosity 1, was used in the second configuration. In a third configuration, the culture broth was used as fluidization promoter; in this case, pure oxygen was bubbled through the reactor.

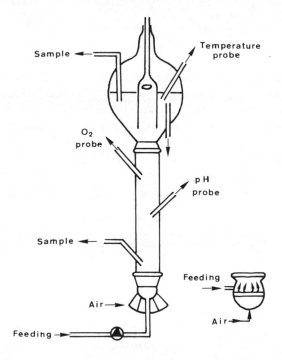

Figure 9 *Bubble column reactor, closed to biomass, for continuous production of monoterpene alcohols by* Ceratocystis moniliformis

23 S. Cohen, H. Chatelain, J.M. Navarro, and J. Crouzet in 'Flavour Science and Technology' ed. M. Martens, G.A. Dalen, and H. Russwurm Jr., J. Wiley, Chichester, 1987, p. 231.

7.1 Hydrodynamic study

The hydrodynamic behaviour of the fluid through the reactor was modelled
from the results obtained from residence time determination (RTD). A preliminary study showed that fluorescein was not retained irreversibly by the
mycelium, and so this colorant was used for residence time studies. The
mean residence time was calculated from the curves giving the variation of
absorbance as a function of time, after fitting of the first part of the curve by
a seventh degree polynomial and extrapolation of the tail by an exponential
function. The use of reduced time θ =t/t.m.s. allows the normalization of
experiment time (Figure 10).

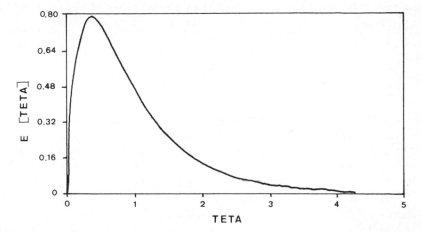

Figure 10 *Residence Time Determination: theorical curve obtained using the bubble column reactor*

The choice of a model for the hydrodynamic behaviour of the reactor was
made by comparison of the experimental and theoretical curves. Due to the
shape of the experimental curves on the one hand, and the configuration of
the reactor used (composed of two parts) on the other, the first model tested
was two continuous flow-stirred tanks in series with two different volumes
V_1 and V_2, α being the volume ratio given by:[24]

$$\theta_{max} = \frac{\alpha.\ln\alpha}{\alpha^2-1}$$

The values obtained for α under different conditions of biomass
concentration, aeration mode, and air flow rate are given in Table 4. The

[24] S. Elmaleh and R. Ben Aim, *Chem. Eng. J. (Lausanne)*, 1975, **9**, 107.

Table 4 *Determination of volume ratio* α *for different biomass content and aeration conditions*

Biomass (g l^{-1})	Aeration mode	Air flow rate (ml min^{-1})	t.m.s. (h)	t (h)	α
0.00	needles	500	2.2	0.11	0.0111
1.26	needles	700	4.3	0.13	0.0059
6.03	needles	600	26.5	3.33	0.1743
0.00	sintered glass	500	2.7	0.14	0.0119
3.25	sintered glass	500	2.2	0.09	0.0086

volumes of the two reactors, expressed as a percentage of the total volume, are given by:

$$\frac{100\alpha}{1 + \alpha} \quad \text{and} \quad \frac{100}{1 + \alpha}$$

When the first configuration of the reactor is operated with water, or at low biomass content, the calculated α values show that the volume of the most important reactor corresponds to more than 98 % of the total volume. Under these conditions, it may be considered that the hydrodynamic behaviour is close to that of one ideal stirred tank. At higher biomass content *i.e.* 6 g l^{-1}, it is possible to consider that the reactor consists of a series of two continuous flow stirred tanks, with volumes respectively equal to 15 % and 85 % of the total volume. In the second configuration, it appears that the reactor may be considered as a quasi-ideal stirred tank.

When the culture medium was used for fluidization, a preliminary study was conducted using a model system composed of alginate balls (Figure 11). From RTD curves obtained under production conditions, it may be concluded that the system is equivalent to a continuous flow-stirred tank. Residence time determinations made using the reactor confirm the existence of channelling, previously found in fluidization experiments performed with alginate balls.

7.2 K_1a Determinations

K_1a values were determined for the first two configurations of the reactor, at different biomass contents and for aeration ratios from 0.44–1.6 v.v.m., using the dynamic method of Taguchi and Humphrey.[25] The concentration of dissolved oxygen, C, when the aeration is re-set after an interruption of air supply is given by:

[25]   H. Taguchi and A.E. Humphrey, *J. Ferm. Technol.*, 1966, **44** 881.

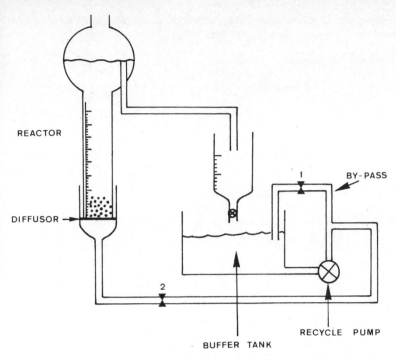

Figure 11 *Reactor used for liquid fluidization studies using alginate balls*

$$C = C^* - \left(\frac{1}{K_1 a} \times \frac{\mathrm{d}c}{\mathrm{d}t} \right) + rx$$

where C^* is the theoretical value of the oxygen concentration in equilibrium with the partial pressure of oxygen, r is the specific rate of oxygen uptake, and x is the biomass expressed in g l^{-1}.

The plot of $C = f(\mathrm{d}c/\mathrm{d}t)$ gives C^* and $K_1 a$. It was assumed that C^* was constant in any part of the reactor due to its small capacity. Likewise we have considered that, because of the low growth rate of the fungus, the biomass content was constant during the determination. As a rule, $K_1 a$ decreases when the aeration ratio increases (Table 5); however, the increased hold-up and agitation is not in agreement with this result. It is necessary to take into account the coalescence of bubbles: because of the flow-rate increase, the number and the size of bubbles are increased, thus the impact between bubbles is facilitated and fusion occurs. The decreased gas–liquid exchange area 'a', resulting from this phenomenon, is not compensated for by the increase of the facility factor K_1.

The important $K_1 a$ values obtained (Table 6), in spite of high biomass content, are relevant to the mycelium behaviour; in this case, the mycelium is well organized in pellets and the viscosity of the culture medium is equivalent to that of water. On the contrary, in some experiments, the $K_1 a$ values

Table 5 K_1a *values obtained for the first configuration of the reactor, for different aeration conditions at constant biomass content*: 3.05 g l^{-1}

Aeration (v.v.m.)	rx (% min^{-1})	K_1a (h^{-1})
0.56	15.3	118
0.80	12.7	107
1.30	12.8	101
1.60	13.8	89

Table 6 K_1a *values for different aeration conditions*[a]

Assay number	Aeration (v.v.m.)	Biomass (g l^{-1})	rx (% min^{-1})	K_1a (h^{-1})
1[b]	0.44	2.67	7.4	312
2[b]	0.56	4.00	7.4	18.7
3[c]	0.80	4.72	7.0	65.6
4[c]	1.0	4.72	7.0	38.3
5[b]	1.2	1.77	9.2	78.6

[a] S. Cohen, H. Chatelain, J.M. Navarro, and J. Crouzet in 'Flavour Science and Technology' ed. M. Martens, G.A. Dalen, and H. Russwurm Jr., J. Wiley, Chichester, 1987, p. 231.
[b] configuration 1 of the reactor.
[c] configuration 2 of the reactor.

are less important (40–80 h^{-1}) for low concentration (1.1–1.9 g l^{-1}) of a filamentous biomass. The values obtained for K_1a under different aeration conditions (Table 6), varying from 19 to 312 h^{-1}, are always important enough to prevent limitation due to gas–liquid mass transfer. Nevertheless, mass transfer in the solid phase may have been insufficient.

7.3   Production of Aroma Compounds

Examples of the production of aroma compounds by *Ceratocystis moniliformis* using the first reactor configuration for different feeding rates (100–200 ml l^{-1}) and different air flow rates (600– 700 ml min^{-1}) are given in Table 7. The aroma production varies quantitatively, but not qualitatively. In all cases, geraniol is the major compound isolated, at about 83 % of the mixture; citronellol and nerol represent, respectively, 12 and 5 % of this product. However in some experiments, less than 2 % of neral and geranial are found. These compounds, sometimes found in aroma compounds produced in batch culture, may be considered either as intermediary products or as products resulting from modifications occurring in the secondary metabolism of the fungus.

Table 7 *Continuous production of monoterpene alcohols by* Ceratocystis moniliformis ATCC 12861 *in the configuration* 1 *of the reactor*[a]

Assay number	Biomass (g l⁻¹)	Mono-terpene alcohols (mg l⁻¹)	Citronellol (%)	Nerol (%)	Geraniol (%)	Productivity (mg l⁻¹h⁻¹)
1	3.13	60.7[b]	12.2	3.8	80.0	1.7
2	–	25.6	12.4	5.5	82.1	2.0
3	6.10	13.9	12.5	4.5	83.0	2.13
4	6.03	26.2	12.7	3.5	83.8	0.74
5	3.13	40.5[b]	11.6	1.2	84.2	3.2

[a] S. Cohen, H. Chatelain, J.M. Navarro, and J. Crouzet in 'Flavour Science and Technology' ed. M. Martens, G.A. Dalen, and H. Russwurm Jr., J. Wiley, Chichester, 1987, p. 231.
[b] about 2 % of neral and geranial are present in these assays.

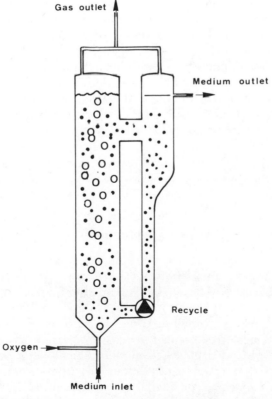

Figure 12 *Bubble column reactor with partial re-cycle of biomass, used for continuous production of monoterpene alcohols by* Ceratocystis moniliformis

The maximum productivity obtained is higher than that found in batch culture: 3.2 against 1.5 mg l^{-1} h^{-1}. This value decreases dramatically when the second configuration of the reactor is used and, more generally, when pellet formation occurs. Organization of the biomass into pellets is favoured by the turbulent air flow used as fluidization promoter, or appears to be a consequence of ageing of the culture. On the other hand, the gaseous stream is responsible for stripping of volatile compounds.

In order to restrict these two phenomena, the use of the culture medium as fluidization promoter was considered; the fluidization was obtained by recycling part of the culture broth through a loop closed to the biomass (Figure 12). In this case, the reactor was fed with pure oxygen. The aroma production under these conditions was low, from 3 to 20 mg l^{-1}. Maximum production was obtained after three days when the density of the fungus was maximum and when the mycelium was not aggregated in pellets. At four days, the density of cells increases, and the pellets appear on the fifth day. These data show clearly the importance of the degree of ageing of the biomass on the production of aroma compounds.

In conclusion, the results obtained show that the productivity is increased when several continuous processes are used for aroma production. However the concentration of aroma components in the culture medium is less important in continuous production than in batch culture. On the other hand it appears, particularly for *Ceratocystis moniliformis*, that ageing of the biomass and pellet formation are responsible for important decreases of aroma production. Under these conditions, the use of a continuous reactor is questionable. From the results obtained for production and productivity, the use of a fed batch reactor is probably a solution, allowing the maintenance of a good productivity with an increased concentration of aroma compounds in the culture broth.

Bioreactors for Industrial Production of Flavours: Use of Enzymes

I.L. Gatfield

HAARMANN & REIMER GMBH, P.O.BOX 12 53, 3450 HOLZMINDEN, GERMANY

1 Introduction

During the last 20–30 years, there has been a significant trend among food manufacturers and consumers in very many countries, away from artificial and towards natural and nature-identical flavours. Recent market surveys have demonstrated that consumers prefer foodstuffs that can be labelled 'natural'. As a direct result, natural flavours are finding wider application in a large variety of food and drink products.

Flavours are almost invariably very complex mixtures of possibly hundreds of individual, flavour-active substances. When natural flavours are to be created, the flavourist has at his disposal quite a broad selection of natural materials with which to work. Such natural materials can be obtained by using traditional techniques such as extraction, distillation, and concentration; but they can also be obtained via biotechnology. For all these categories, the starting materials have to be of natural origin as is stipulated in the definition of a natural flavour according to the US Code of Federal Regulations:

> ... the essential oil, oleoresin, essence or extractive, protein hydrolysate, distillate, of any product of roasting, heating, or enzymolysis, which contains the flavouring constituents derived from a spice, fruit juice, vegetable or vegetable juice, edible yeast, herb, bud, bark, root, leaf or similar plant material, meat, seafood, poultry, eggs, dairy products, or fermentation products thereof, whose significant function in food is flavouring rather than nutrition.

There are two possible approaches for the biotechnological manufacture of natural flavour materials, depending upon the way they are produced, namely via microbial fermentation or via enzymatic methods. As early as

1956, Hewitt and co-workers[1] reported the flavour lost in food processing could in some cases be restored to the food product through the addition of an appropriate enzyme preparation. A blanched, dehydrated sample of fresh watercress was used which was flavourless when rehydrated. When a tasteless, odourless enzyme preparation from white mustard was added to the rehydrated product, the typical odour and taste of watercress was rapidly regenerated. The work was extended to other vegetables including cabbage, string beans, onions, and even to fruit such as raspberries.[2] This work elegantly demonstrated the importance of enzymes in the flavour development of food products.

Enzymes, either in a crude or a purified form, have traditionally been widely used as processing aids in the food industry. As Hewitt's pioneering work showed, biological processes based on enzymes can contribute to the development of flavour in certain foodstuffs. If this is to be utilized on a larger scale, it would seem that two distinct strategies are possible which would result in either complex, multicomponent flavour systems or individual flavour compounds.

2 Production of Complex Mixtures of Flavour-active Compounds

During the ripening process of cheese, a number of enzymatic reactions proceed at the same time, thereby giving rise to an extremely dynamic system. Some of the reactions are catalysed by microbial enzymes and others by those enzymes present in the milk. Milk fat, protein, and carbohydrate are degraded to varying extents thereby giving rise to very complex mixtures of compounds, some of which contribute to the characteristic cheese flavour. Among the multitude of reactions catalysed by enzymes are those involving the hydrolysis of appropriate substrates such as fats, leading to the production of intensely flavoured enzyme-modified products.

2.1 Fatty Acid Mixtures

One of the very first flavours produced on a large scale with the aid of enzymes, was a product known as lipolysed milk fat.[3] The original process involved subjecting cream to a controlled enzymatic hydrolysis using lipases.[4, 5] The composition of the fatty acids liberated depends upon the specificity of the lipases used. Certain lipases exhibit a high degree of

[1] E.J. Hewitt, D.A.M. MacKay, K.S. Konigsbacher, and T. Hasselstrom, *Food Technol.*, 1956, **10**, 487.
[2] E.J. Hewitt, T. Hasselstrom, D.A.M. MacKay, and K.S. Konigsbacher, US P. 2 924 541.
[3] D.J. Pangier, US P. 3 469 993.
[4] A. Kilare, *Process Biochem.*, 1985, **20**, 35.
[5] J. Dziezak, *Food Technol.*, 1986, **40** (4), 114.

specificity towards the flavoursome short chain fatty acids[6, 7] as is shown in Figure 1 for the lipase from *Mucor miehei*. Other lipases preferably liberate long chain fatty acids and others still do not display any particular preference.[4] Lipases also exhibit an additional type of specificity namely towards the position of the fatty acid residue in the triglyceride molecule. The lipolytic treatment of cream has to be carefully controlled since the long chain fatty acids can give rise to unpleasant, soapy flavour notes, especially at higher concentrations. Lipolysed milk fat products find wide application, for example for the enhancement of butter-like flavours[5] and flavour development in milk chocolate.[8, 9, 10]

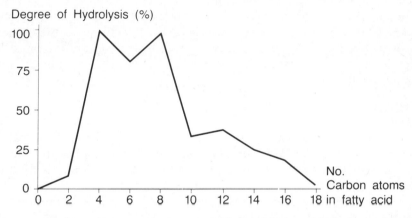

Figure 1 *Effect of chain length of fatty acid upon degree of hydrolysis of synthetic triglycerides by esterase-lipase from* Mucor miehei

Further development of this type of technology led to a new type of product known as enzyme-modified cheese (EMC). Enzyme-modified cheeses are products obtained by a controlled proteolytic and/or lipolytic enzyme treatment of a previously manufactured traditional natural cheese.[11,12] Such a product has significantly increased amounts of proteolytically and/or lipolytically derived flavour compounds characteristic of the cheese which was treated. After thermal inactivation of the enzymes, a product is obtained which can easily have a flavour intensity of 20 times that of the mature cheese. EMC products have become very important materials during the last 20 years or so. It is very difficult to estimate the size of the market since there are many captive producers.

[6] G.J. Moskowitz, R. Cassaigne, I.R. West, T. Shen, and L.I. Feldman, *J. Agric. Food Chem.*, 1977, **25**, 1146.
[7] T. Kanisawa, Y. Yamaguchi, and S. Hattori, *Nippon Shokuhin Kogyo Gakkai-Shi*, 1982, **29**, 693.
[8] A.W. Anti, *Food Proc. Dev.*, 1969, **3**, 17.
[9] T. Hagashi, *Kobunshi*, 1967, **16**, 1220.
[10] Tanabe Seiyaku Ltd. Product Information Bulletin on Lipase RH, Osaka, Japan.
[11] V.K. Sood and F.V. Kosikowski, *J. Dairy Sci.*, 1979, **62**, 1865.
[12] G.J. Moskowitz and S.S. Noelk, *J. Dairy Sci.*, 1987, **70**, 1761.

3   Production of Individual Flavour-active Compounds

During the last ten years or so, a number of enzymatic methods have been developed for the production of specific flavour-active compounds. Enzyme-catalysed syntheses are typically performed at more moderate values of temperature, pressure, and pH than are similar chemical syntheses. Enzymes may also display remarkable enantio- and regiospecificity. However, enzymatic syntheses do suffer from certain drawbacks because the reactions are comparatively slow, the enzyme may be unstable or require cofactors, yields can be low and the enzyme or cofactor can be very expensive. Nevertheless, enzymatic processes have been developed for the production of a number of flavour compounds.

The vast majority of commercially available enzymes are hydrolytic ones. Of the 75 000 tonnes of enzymes produced in 1985, more than 80 % were hydrolytic.[4] It is therefore not surprising that hydrolases are the ones that have been evaluated most as regards their ability to produce individual flavour materials of interest.

3.1   Fatty Acids

Short chain fatty acids exhibit a very intense flavour. Typical representatives of this class of materials are propionic, butyric, caproic, and caprylic acids. One source of fatty acids with an even number of carbon atoms is milk fat. There is a product on the market called butter acids which contains the naturally occurring fatty acids from butter. Such a product is traditionally made by the saponification of butter fat. Since this is a non-specific process all the fatty acids originally present in triglyceride form in the butter, are liberated. Butyric acid is present to the extent of approximately 3–4 % and this is the highest concentration of the lower aliphatic acids.

The lipolytic hydrolysis of butter fat represents an alternative procedure which also gives rise to a natural product. Homogenized butter fat undergoes rapid lipolysis when treated for example with pancreatin under constant conditions of pH.[16] After an incubation period of 5–6 hours the enzymatic reaction comes to a standstill. The liberated fatty acids can be isolated by steam distillation and the individual acids then further purified by fine distillation. In this way the short chain fatty acids such as butyric, caproic, caprylic, and capric acid can be obtained in a pure form.

3.2   Esters

Esters are very important flavour compounds and occur widespread in nature in a great variety of foodstuffs. Esters typically exhibit fruity flavour notes. There are many reports in the literature which show that esters are biosynthetic products of many mirco-organisms. Fruity flavour defects sometimes found in Cheddar cheese and beer are due to the presence of esters, for example ethyl butyrate.[13]

[13] A. Hosono and J.A. Elliott, *J. Dairy Sci.*, 1975, **57**, 1432.

We have been actively engaged for many years in the application of enzyme systems for genuine preparative purposes. Thus, the synthetic properties of the lipase obtained from the mould *Mucor miehei* has been studied in considerable detail.[14, 15] This enzyme is used in the manufacture of cheese. It could be shown that this enzyme displays a remarkable synthetic activity not only in non-aqueous systems but also in such systems where no organic material other than the acid and alcohol components is present. In other words, the enzyme is very active not only in non-aqueous but also in solvent-free systems.

It is generally recognized in the meantime that manipulation of the reaction media can result in a shift of the thermodynamic equilibrium to favour the synthetic process of hydrolytic enzymes. Typical esterification profiles obtained for our system consisting of equimolar quantities of acid and alcohol are shown in Figure 2. Under the non-aqueous conditions prevailing,

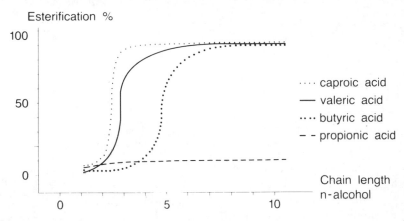

Figure 2 *Effect of chain length of carboxylic acid upon the degree of esterification obtained with n-alcohols using the lipase from* Mucor miehei

the enzyme exhibits a distinct specificity as regards the chain lengths of both the acid and the alcohol components. Generally speaking, it seems that the longer the chain lengths of the alcohol and acid components, the higher the degree of esterification achieved. Esters of propionic and acetic acid are more difficult to synthesize this way. However, good yields can be obtained in the presence of a second long chain fatty acid, *e.g.* oleic acid.[16] It could be shown that in such systems the alkyl oleate was the first product formed and this underwent enzymatic transesterification to form the alkyl propionate or alkyl acetate in good yields.

[14] I.L. Gatfield and T. Sand, German P. 3 108 927.
[15] I.L. Gatfield, *Ann. N.Y. Acad. Sci.*, 1984, **434**, 569.
[16] I.L. Gatfield, unpublished results.

The enzyme is usually very stable and in very many cases it can be re-used a number of times without significant reduction in its ability to synthesize esters. What has often been detected however, is a drastic reduction in the lipolytic activity of the enzyme.[16] The data obtained for a typical experiment are shown in Table 1. Thus the enzyme loses approximately 90 % of its lipolytic activity during the first cycle, and the 10 % residual activity remains constant during the next five cycles. The loss of lipolytic activity does not affect the synthetic capabilities of the enzyme, since the degree of esterification remains constant at around 85 % during the first five cycles. In the sixth cycle the synthetic ability of the enzyme declines considerably.

Table 1   *Data from the synthesis of isoamyl isovalerate using the lipase from* Mucor miehei

Cycle	Degree of Esterification (%)	Lipolytic Activity (%)
0		100.0
1	85.2	9.8
2	88.0	9.5
3	85.0	10.0
4	86.4	9.0
5	82.1	8.5
6	53.7	8.0

These findings suggest that the ester-synthesizing activity of this enzyme has little to do with its lipolytic activity. The latter is more or less completely lost early on during ester synthesis. Furthermore, additional synthetic cycles can be carried out just as efficiently by using the enzyme which shows only very little lipolytic activity. This is the case when both straight and branch chain acids and alcohols are used as substrates.

Gel permeation chromatography of the air-dried enzyme residue from another preparative batch was performed on Sephadex G-10 and the higher molecular weight fraction was freeze dried. The same was done with the fresh enzyme preparation which had not been used for ester synthesis. CD spectra of these two higher molecular weight fractions were measured in water as solvent. Although the fractions are by no means pure protein, the spectra indicate that a conformational change in the protein structure takes place as a result of the enzymatic esterification reaction. Thus, the position of the negative Cotton effect has been shifted to lower wavelengths and has become somewhat weaker (Figure 3). This conformational change is presumably severe enough to almost eliminate the lipolytic activity but does not have such an adverse effect upon the ester-synthesizing activity.

It is well known that water-miscible organic solvents such as ethanol can disrupt the hydrophilic interactions between the protein molecule and its hydration layer thereby allowing the organic liquid to penetrate the

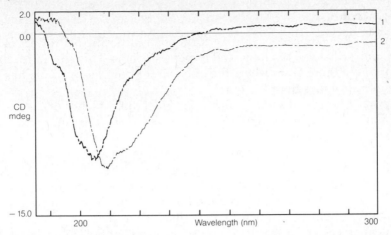

Figure 3 *CD-Spectra of aqueous solutions of the two higher molecular weight fractions from:* 1. *enzyme residue after esterification reaction;* 2. *unused lipase*

hydration layer and, possibly, denature the protein structure.[17] This is possibly the explanation for these experimental findings.

Quite a number of other lipases have been evaluated as to their suitability for synthesizing esters in both aqueous and non-aqueous systems. Thus, the immobilized lipase from *Candida cylindracea* was used to synthesize ethyl butyrate.[18] The enzymatic activity was maintained through several cycles of enzyme reuse by simple hydration of the enzyme column between uses. The lipase from *Aspergillus niger* and pancreatin can also be used to synthesize ethyl butyrate.[19] Esters of terpene alcohols are also accessible via enzymatic synthesis. In one report some 50 different hydrolases were examined for esterification activity towards citronellol.[20] Very many of them exhibited high citronellol esterification activity in organic solvents.

The lipase from *Candida cylindracea* is also capable of synthesizing an ethyl ester mixture from butter fat.[21] It was observed that this lipase effectively converted the free fatty acids released from the butter fat by the lipase into the corresponding ethyl esters in the presence of ethanol. Thus, unhydrolysed butter fat was treated with the lipase in the presence of 10–12 % ethanol. Ethyl ester profiles at various stages of esterification showed that butyric and capric acids were esterified more rapidly than others. In this system both hydrolysis and esterification take place at the same time. Presumably, the natural water content of the butter fat is sufficient to promote the hydrolysis reaction and is not too high to cause inhibition of the esterification reaction.

[17] J.C. Duarte, in 'Perspectives in Biotechnology Vol. 128', ed. J.C. Duarte, C.J. Archer, A.T. Bull, and G. Holt, Plenum Press, New York, 1987, p. 23.
[18] B. Gillies, H. Yamazaki, and D.W. Armstrong, in 'Biocatalysts in Organic Media', ed. C. Laane, J. Tramper, and M.D. Lilly, Elsevier, Amsterdam, 1987, p. 227.
[19] F.W. Welsh and R.E. Williams, *Enzyme Microb. Technol.*, 1990, **12**, 743.
[20] T. Kawamoto, K. Sonomoto, and A. Tanaka, *Biocatalysis*, 1987, **1**, 137.
[21] T. Kanisawa, *Nippon Shokuhin Kogyo Gakkai-Shi*, 1983, **30**, 572.

3.3   Alcohols

Lipases have also been used to perform enantiospecific hydrolyses to yield pure, optically active aliphatic and terpene alcohols. This type of reaction is important in those cases where the one chiral form of a molecule has more desirable attributes than the other.

A prime example of such an alcohol is *l*-menthol which is one of the most important terpene alcohols and is widely used in the fragrance and flavour industry. *l*-Menthol is the isomer which occurs most widely in nature and is the main component of peppermint oil. *l*-Menthol has a characteristic peppermint odour and also exerts a cooling effect. The other isomers do not possess this cooling effect and are, therefore, not considered to be 'refreshing'. *d, l*-Menthol occupies an intermediate position; the cooling effect of the *l*-menthol present is distinctly perceptible. There are quite a number of chemical processes for the resolution of the two menthol isomers and this is the way nature-identical *l*-menthol is produced in the industry. Nevertheless, biochemical resolution methods have also been developed.

A number of research groups have directed their attention towards the application of the fact that many microbial lipases preferentially hydrolyse *l*-menthyl esters while leaving the *d*-menthyl esters untouched. This is shown schematically in Figure 4. Patents have been filed by several companies covering a wide range of micro-organisms and their constituent lipases and/or esterases. Selected species of *Penicillium*, *Rhizopus*, *Bacillus*, and *Trichoderma* for example, perform this asymmetric hydrolysis reaction on the acetate, propionate, and caproate esters of *d, l*-menthol.[22] The asymmetric cleavage of *d, l*-menthyl succinate has also been reported.[23, 24]

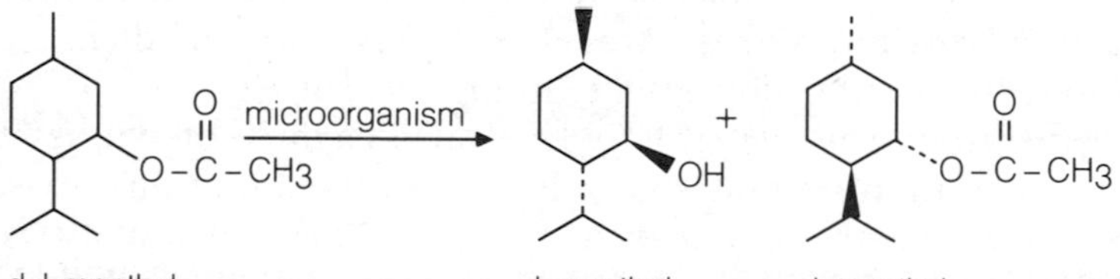

d, l-menthyl   l-menthol   d-menthyl
acetate   acetate

Figure 4   *Preferential hydrolysis of* d,l-*menthyl acetate by microbial lipases*

The opposite approach has been published in a Japanese patent[25] which describes the enzyme-catalysed preferential esterification of *l*-menthol with certain fatty acids using the lipase from *Candida lipolytica*. An extension of this work has been published recently in which the specificity of the enzyme synthesis was studied using lauric acid and various α-substituted *d, l*-

[22] T. Moroe, S. Hattori, A. Kamatsu, and Y. Yamaguchi, US P. 3 607 561.
[23] T. Oritani and Y. Yamashita, *Agric. Food Chem.*, 1973, **37**, 1697.
[24] S. Fukui, *Appl. Microbiol. Biotechnol.*, 1981, **22**, 199.
[25] Meito Sangyo Co. Ltd., Jap. P. 126 375 and J 5-2051-095.

cyclohexanols in apolar solvents.[26] In the case of *d, l*-menthol it could be shown that an efficient optical resolution could be achieved. Thus, after a reaction period of eight hours, a degree of esterification of 45 % was obtained whereby *d*-menthyl laurate was the major product. The remaining unreacted alcohol was mainly *l*-menthol which showed an enantiomeric excess of 90 %.

Despite the many positive results found in the literature, it would seem that the biotechnological approach towards the optical resolution of *d, l*-menthol still cannot compete with the traditional method which entails the selective crystallization of suitable derivatives.[27]

3.4 Lactones

Lactones are generally very pleasant, potent flavour materials which are widely distributed in Nature. Lactones have been isolated from all major classes of food including fruits, vegetables, nuts, meat, milk products, and baked goods.[28] The microbiological synthesis of lactones attracted the attention of scientists in the 1960s. Most systems involved the use of viable microorganisms which are not the topic of this chapter. However, for the sake of completion, two of the fermentation procedures should be mentioned briefly.

An efficient reduction of the corresponding keto-acids to the optically active hydroxy-acids was achieved by using yeasts, moulds, and bacteria. The hydroxy-acids were then converted chemically into the corresponding lactones.[29] Similarly, a procedure was described by Okui[30] in 1963 in which ricinoleic acid was degraded via β-oxidation by various *Candida* species to form 4-hydroxydecanoic acid which can be cyclized to form γ-decalactone. This is shown schematically in Figure 5.

Lactones can also by synthesized enzymatically from the corresponding hydroxycarboxylic acids. During the evaluation of the lipase from *Mucor miehei*, we also investigated this possibility.[14] Thus γ-butyrolactone was obtained in good yield from 4-hydroxybutyric acid and 15-hydroxypentadecanoic acid was converted into the macrocyclic pentadecanolide (Figure 6). The latter is an interesting fragrance material and displays a typical musk note. An important aspect of this intramolecular cyclization reaction is that it should be performed at high dilution (Figure 7). If this is not heeded, polymer formation occurs through intermolecular condensation reactions.[16]

These observations have been confirmed by Yamada,[31] who reported the

26 G. Langrand, M.Secchi, G. Buono, J. Baratti, and C. Triantaphylides, *Tetrahedron Lett.*, 1985, **26**, 1857.
27 J. Fleisher, K. Bauer, and R. Hopp, German P. 2 109 456.
28 J.A. Maga, *CRC Crit.Rev. in Food Sci. Nutr.*, 1976, **10**, 1.
29 G.T. Muys, B. van der Ven, and A.P. De Jonge, *Appl. Microbiol.*, 1963, **11**, 389.
30 S. Okui, M. Uchiyama, and M. Mizugaki, *J. Biochem.*, 1963, **54**, 536.
31 A. Makita, T. Nihira, and Y. Yamada, *Tetrahedron Lett.*, 1987, **28**, 805.

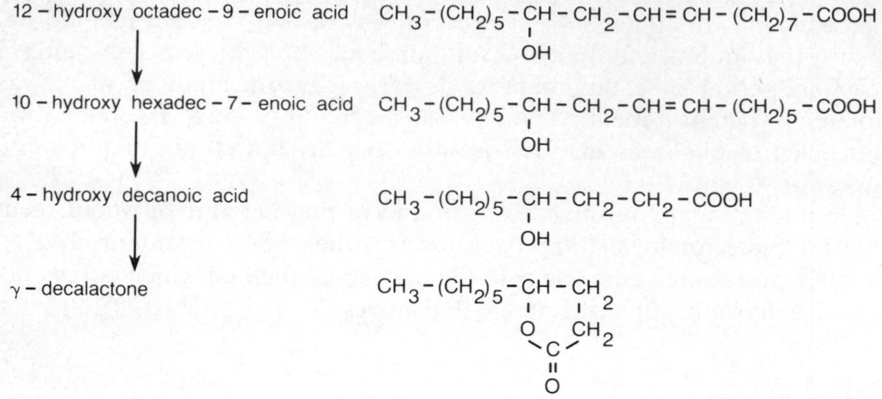

12 – hydroxy octadec – 9 – enoic acid $\quad CH_3-(CH_2)_5-\underset{\underset{OH}{|}}{CH}-CH_2-CH=CH-(CH_2)_7-COOH$

$\downarrow$

10 – hydroxy hexadec – 7 – enoic acid $\quad CH_3-(CH_2)_5-\underset{\underset{OH}{|}}{CH}-CH_2-CH=CH-(CH_2)_5-COOH$

$\downarrow$

4 – hydroxy decanoic acid $\qquad\qquad CH_3-(CH_2)_5-\underset{\underset{OH}{|}}{CH}-CH_2-CH_2-COOH$

$\downarrow$

γ – decalactone $\qquad\qquad\qquad\quad CH_3-(CH_2)_5-CH-CH_2$

Figure 5 *Oxidative degradation of ricinoleic acid by* Candida lipolytica

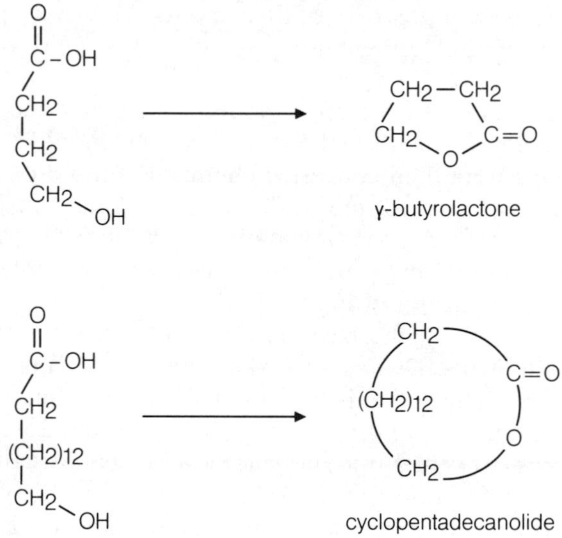

γ-butyrolactone

cyclopentadecanolide

Figure 6 *Lactone synthesis from the corresponding hydroxy-acids using the lipase from Mucor miehei*

biocatalytic lactonization of ω-hydroxy-acid methyl esters in highly diluted solutions. Also Gutman *et al.*,[32] found that porcine pancreatic lipase in anhydrous organic solvents catalysed the lactonization of a number of esters of ω-hydroxy-acids with high degrees of enantiomeric specificity.

[32] A.L. Gutman, K. Znobi, and A. Boltansky, *Tetrahedron Lett.*, 1987, **28**, 3861.

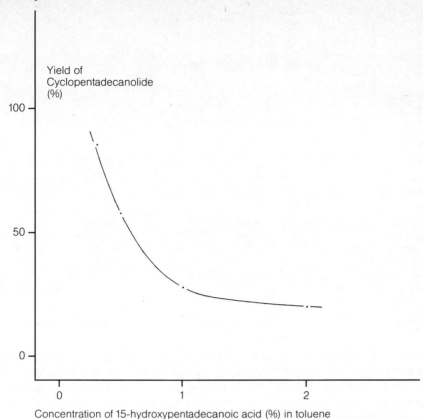

Yield of
Cyclopentadecanolide
(%)

Concentration of 15-hydroxypentadecanoic acid (%) in toluene

Figure 7 *Effect of dilution upon the yield of cyclopentadecanolide during the enzymatic cyclization of* 15-*hydroxypentanoic acid*

3.5 Aldehydes and Ketones

There are very many aldehydes and ketones which are important ingredients of natural flavours. It is, for example, almost impossible to imagine a yoghurt or orange flavour which does not contain a certain amount of acetaldehyde. This material is naturally present in natural yoghurt and natural orange juice and determines to a considerable extent the overall flavour of both products.

The same goes for higher aldehydes such as *cis*-3-hexenal which contributes together with its corresponding alcohol towards the flavour of many types of ripe fruit. The *cis*-3-hexenal arises through the action of lipoxygenase on the unsaturated fatty acid, linolenic acid followed by the action of hydroperoxide lyase (Figure 8). Linoleic acid undergoes the same enzymatic transformation and gives rise to hexanal. This mechanism has been known for quite a long time.[33]

[33] F. Drawert, W. Heimann, R. Emberger, and R. Tressl, *Liebigs Ann. Chem.*, 1966, **694**, 200.

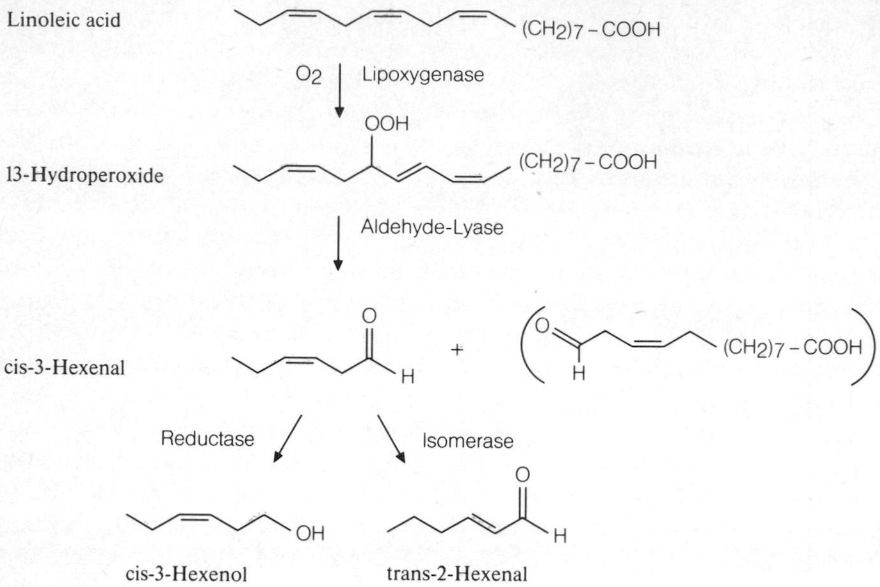

Linoleic acid

O_2 | Lipoxygenase

OOH

13-Hydroperoxide

| Aldehyde-Lyase

cis-3-Hexenal

Reductase / \ Isomerase

cis-3-Hexenol   trans-2-Hexenal

Figure 8   *Enzymatic formation of aldehydes and alcohols from unsaturated fatty acids*

Systems for the preparative scale production of such 'green' flavour components via the oxidation of fatty acids using isolated enzymic systems, have not been described so far. The likely reason for this is the lack of availability of the lyase enzyme. Efforts to obtain these types of flavour compounds seem to be restricted to the use of endogenous enzymes present in the plant tissues.

Thus, a process has recently been patented for the production of a green leaf essence.[34] The process comprises homogenizing strawberry leaves in water to form a slurry. To this slurry is added between 1 and 10 mM of linolenic acid which was obtained from linseed oil. The slurry is stirred for about 24 hours and then the volatile materials are isolated. The major components are *cis*-3-hexenal, *trans*-2-hexenal, and *cis*-3-hexenol which together make up approximately 88 % of the total volatiles.

A similar process for the preparation of green flavour compounds has been patented in which raw soybeans are used as the source of the necessary enzyme systems.[35] The yields of volatile materials obtained this way are typically very small. Thus, approximately 7300 mg of volatile material was isolated from some 15 kg of fresh strawberry leaves.[34]

A very similar system has been patented for the manufacture of 1-octen-3-ol which is the character impact flavour compound in edible mushrooms.[36] A few years ago, Grosch *et al.*,[37] have shown that mushrooms contain a

[34] S. Goers, P. Ghossi, J.T. Patterson, and C.L. Young, US P. 4 806 379.
[35] T. Kanisawa and H. Itoh, US P. 4 769 243.
[36] F. Schindler, German P. 3 708 932.
[37] M. Wurzenberger and W. Grosch, *Biochim. Biophys. Acta.*, 1984, **795**, 163.

lipoxygenase which produces the 10-hydroperoxide of linoleic acid. This hydroperoxide undergoes cleavage by the corresponding hydroperoxide lyase to form 1-octen-3-ol.

An alternative approach for the production of aldehydes would be via an enzymatic oxidation of their respective alcohols. Only two oxidoreductases have been seriously examined for the production of this type of flavour material. These enzymes are alcohol dehydrogenase and alcohol oxidase and both oxidize aliphatic alcohols to their respective aldehydes. The enzymes necessary for such an oxidation step almost invariably require cofactors. For many cofactors, their initial cost prevents one-time usage on a large scale. This cost necessitates a high degree of cofactor recycling if the reaction being catalysed is to approach the economy of scale typical of hydrolytic enzymes.

Various methods are used for cofactor regeneration. A procedure was patented in 1984 for the conversion of ethanol to acetaldehyde[38] using the enzyme alcohol dehydrogenase (ADH). This process is complex (Figure 9) due to the necessity of cofactor regeneration. Specifically, the process involves the use of nicotinamide adenine dinucleotide cofactor, which during the course of the reaction is reduced and then regenerated by light-catalysed oxidation with flavin mononucleotide (FMN). The reduced flavin mononucleotide ($FMNH_2$) is reconverted to FMN by oxidation with molec-

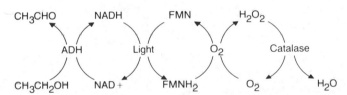

Figure 9 *Enzymatic oxidation of ethanol to acetaldehyde according to* US P. 4 481 292

ular oxygen. The by-product of this reconversion is hydrogen peroxide which is in turn decomposed, by the action of the enzyme catalase, to oxygen and water. Conversion rates in the range of 10–20 % are typically obtained and in a continuous batch reactor system, the concentration of acetaldehyde achieved is some 2.5 g l^{-1} after a period of nine hours.

Horse liver alcohol dehydrogenase can convert the monoterpene alcohol geraniol to its corresponding aldehyde, geranial.[39] Here a biphasic system with hexane as the organic solvent was used as the reaction medium. The substrate geraniol, diffused freely into the aqueous phase, underwent the enzymatic conversion to geranial and the product diffused back into the organic phase. A coupled substrate method was used to regenerate the NAD$^+$ cofactor utilizing the reduction of acetaldehyde to ethanol. This model system was also used successfully for the production of citronellal,

[38] W.R. Raymond, US P. 4 481 292.
[39] M.D. Legoy, H.S. Kim, and D. Thomas, *Process Biochem.*, 1985, **20**, 145.

hexanal, and 3-phenylpropanal. Similar work has been conducted on the bioconversion of *trans*-cinnamaldehyde to *trans*-cinnamyl alcohol. This reduction was carried out by yeast alcohol dehydrogenase.[40]

While ADH can operate in both directions, alcohol oxidase which is found in methylotrophic yeasts, operates in one direction only and converts alcohols to aldehydes.[41] A good example for this type of enzyme is that obtained from *Pichia pastoris*. The enzyme system from this organism is able to oxidize a wide range of low molecular weight primary alcohols[42] and also aromatic alcohols.[43]

3.6 Sulphur-containing Materials

Sulphur-containing compounds are found in numerous foods as part of their natural flavour complex and are formed as a result of thermal or enzymatic processes. For example, the pungent isothiocyanates found in mustard are formed by the hydrolytic release of these compounds from thioglucoside or glucosinolate precursors by the enzyme myrosinase (Figure 10). A system has been described[44] for the use of immoblized myrosinase for the continuous production of horseradish flavour. The flavours of onions, leeks, and garlic are formed as a result of the conversion of the precursors *S*-alkylcysteine sulphoxides by the enzyme alliinase.

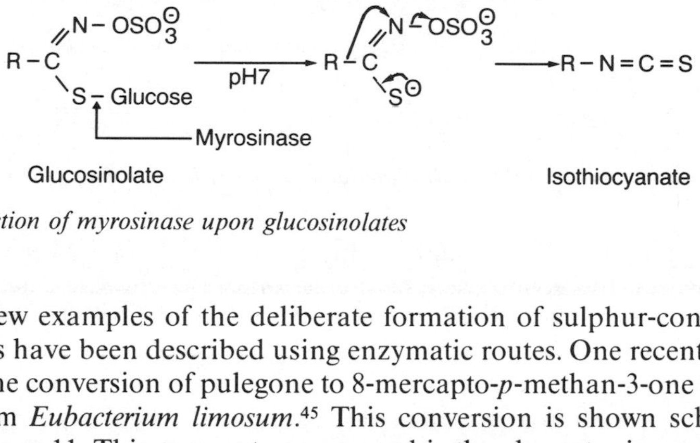

Figure 10 *Action of myrosinase upon glucosinolates*

Only a few examples of the deliberate formation of sulphur-containing compounds have been described using enzymatic routes. One recent patent describes the conversion of pulegone to 8-mercapto-*p*-methan-3-one using a β-lyase from *Eubacterium limosum*.[45] This conversion is shown schematically in Figure 11. This mercapto compound is the character impact material of the blackcurrant and is a natural constituent of buchu leaf oil.

[40] W.R. Bowen, N. Lambert, S.Y.R. Pugh, and F. Taylor, *J. Chem. Technol. Biotechnol.*, 1986, **36**, 267.
[41] T.R. Hopkins and F. Müller, in 'Microbial Growth on C₁ Compounds', ed. H.W. van Verseveld, and J.A. Duine, Kluwer, Dordrecht, 1987, p.150.
[42] W.D. Murray, S.J.B. Duff, and P.H. Lanthier, US P. 4 871 669.
[43] S.J.B. Duff and W.D. Murray, *Biotechnol. Bioeng.*, 1989, **34**, 153.
[44] I.L. Gatfield, G. Schmidt-Kastner, and T. Sand, German P. 3 312 214.
[45] A. Kerkenaar, D.J. Schmedding, and J. Berg, Eur. P. Applic., 277 688.

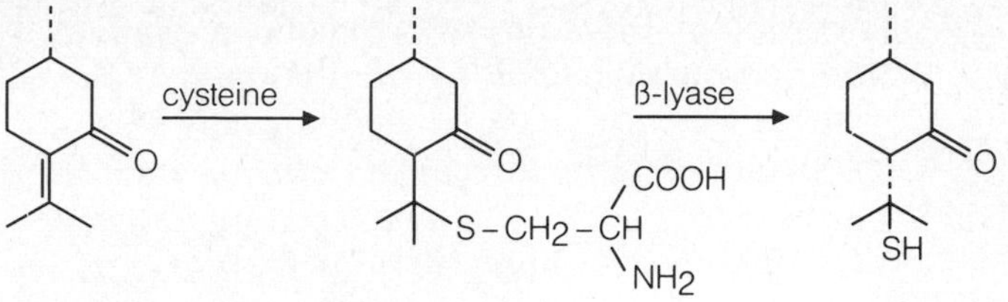

pulegone S-cysteinyl pulegone 8-mercapto-p-menthan-3-one

Figure 11 *Conversion of pulegone to* 8-*mercapto*-p-*menthan*-3-*one using the* β-*lyase from* Eubacterium limosum

4 Future Prospects

As described in this chapter there are quite a number of enzymatic and microbiological processes which are already being used to produce materials of interest to the industry. There is great potential for the further development of such processes for flavour biosynthesis. The use of non-aqueous and biphasic systems will become more important. It is possible that further advances will be made when enzymes are used under unusual conditions. Care must be taken to ensure that the systems developed are appropriate for the final use of such materials in foodstuffs. This type of technology could also be used to develop perfumery materials.

Downstream Processing: Concentration and Isolation Techniques in Industry

H.J. Takken, N.M. Groesbeek, and R. Roos

QUEST INTERNATIONAL, P.O. BOX 2, 1400 CA BUSSUM, THE NETH-ERLANDS

1 Introduction

Traditionally, flavours are composed of ingredients derived from plant material, micro-organisms, protein, 'Maillard' chemistry-based products, and synthetic chemicals (Figure 1).

Due to the present trends in the market place toward more natural products, there is a clear need for new natural sources for flavour ingredients (Figure 2). It is because of this trend that biotechnology has gained much importance for flavour production.

Microbiological flavour formation has a long tradition in food processing (Figure 3).

At present a broad spectrum of flavour ingredients is available from biotechnological processes (Figure 4).

Biotechnology combines biochemistry, microbiology, and process technologies, in which downstream processing plays and important role.

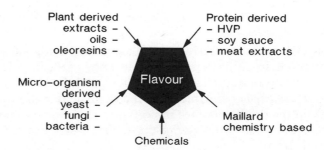

Figure 1 *Traditional flavour ingredient sources*

- healthiness
- convenience
- improved quality
- naturalness

➡ new natural ingredients
⬆
BIOTECHNOLOGY

Figure 2   *Trends in the market*

Purpose   : ● preservation
● texturizing
● improvement in palatability
● flavour formation

Examples : ● beer, wine
● cheese, butter, yoghurt
● meat
● soy sauce

Figure 3   *Traditional role of fermentation in foods*

● alcohols	● complex building blocks
● acids	
● esters	● amino acids
● aldehydes	● peptides
● ketones	● nucleotides
● lactones	
● N and S	
compounds	

Figure 4   *Flavour ingredients by biotechnology*

2   Downstream Processing

Downstream processing (DSP) is a collective name for a broad variety of technologies used in isolation, concentration, and purification of products from a fermentation process.

Several reasons make it necessary to isolate a flavour ingredient from its fermentation broth. In most cases the concentration of the ingredient will be too low to use the broth as such as an ingredient, and for many applications an aqueous product is not suitable. Also the ingredient might not be sufficiently stable in the broth, but most importantly, the olfactive quality of the product will, in many cases, deteriorate due to the presence of many fermentation by-products.

The required downstream processing steps can add considerably to the final cost of a product. From an Arthur D. Little report,[1] a graph is abstracted indicating the relationship between the concentration of a product in a broth and the selling price. Without any doubt, DSP is responsible for a major part of the integral cost of the products (Figure 5). For bulk fermentation products, contributions of 20–50 % to the costs are reported.[2]

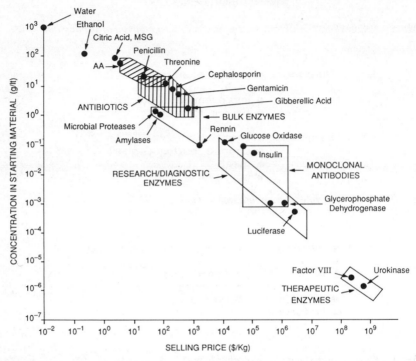

Figure 5   *Downstream processing – major cost factor*

In the total process from raw material to flavour ingredient, DSP is essential to arrive at the right organoleptic quality of the product (Figure 6). It can, however, be equally important to pay considerable attention to the quality of the precursors that will be transformed in the fermentation process. By-products accompanying the precursor can have an influence on the micro-organism—*e.g.*, lead to retarded growth of the organism or slow down the biotransformation process—or may influence the quality of the final product or complicate the downstream processing.

To exemplify the importance of upstream processing, the isolation and purification of 11-hydroxypalmitic acid from the precursor plant material is

[1] 'Biotechnology and Dutch Industry', Arthur D. Little, Report to Dutch Ministry of Economic Affairs, 1990.
[2] A.J. Hacking, in 'Food Biotechnology', ed. R.D. King and P.S.J. Cheetham, Elsevier, London, 1988, p. 33.

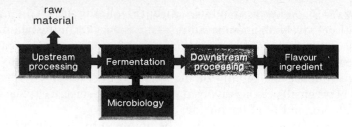

Figure 6   *Position of downstream processing in flavour ingredient production*

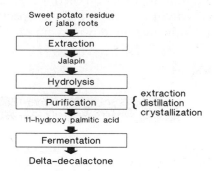

Figure 7   δ-*Decalactone precursor*

shown in Figure 7. Experience has shown that a pure hydroxy-acid is required for an efficient biotransformation into δ-decalactone.

The whole sequence of upstreaming–fermentation–downstreaming should be considered as an integral process. Since ultimately the product will play a role in the development of flavours, the organoleptic quality of the product is of major importance. This quality should not only be of high standard but it should, in particular, be constant, thus requiring a very reproducible process. However, a fermentation process inherently has a number of aspects which are extremely difficult, if not impossible, to standardize. Both the living organisms and the composition of the nutrients—often derived from living organisms—can give rise to a non-reproducible process (Figure 8). We have observed that different batches of a peptone can lead to severe problems in the extraction process and to different by-product composition.

● broth composition : – yeast autolysates
                              – protein hydrolysates

● sterilization

● living micro–organisms

Figure 8   *Possible causes of irreproducibility*

Sterilization conditions may also affect the quality of the final product. Areas with relatively high temperatures—*e.g.*, the jacket or the heating coil—may give rise to Maillard-type reactions causing unacceptable organoleptic properties or browning of the final product. Because of the many aspects that can influence the quality of the final product, close co-operation is required between all persons involved in the research stage, technical development, scale-up, and final flavour development.

Flavour ingredients can sometimes be obtained by *de novo* bioformation by micro-organisms. However, in many cases a biotransformation of a well-chosen precursor is a more efficient approach. Both the precursor and the product may have an inhibiting effect on the micro-organisms.

To increase productivity in such processes special approaches are required, which greatly influence the downstream process[3, 4] (Figure 9).

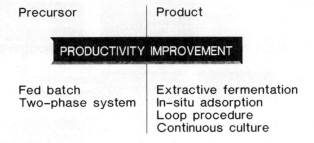

Figure 9   *Inhibition*

Possible approaches are:
(*a*)   Extractive fermentation, whereby a suitable water-immiscible solvent is added to the fermentation broth in order to reduce the concentration of the product in the water phase.
(*b*)   Addition of adsorbents. The adsorbent should have a good affinity for the inhibiting product, but no affinity for the polar media components and it should not affect viability of the micro-organism.
(*c*)   An effective but rather expensive solution could be to build in a loop procedure whereby part of the fermentation broth is continuously pumped over a column with an appropriate adsorbent (Figure 10).

3   Factors Influencing DSP Design

Regulations set by Governmental Authorities, Industrial Associations, and the Company involved on 'naturalness', safety, GMP, and environmental aspects have a major influence on DSP (Figure 11). The regulation for the use of extraction solvents, as given by the EEC directive,[5] limits our

[3]   S.R. Roffler, H.W. Blanch, C.R. Wilke, *Trends in Biotechnology*, 1984, **2** (5), 129.
[4]   B.M. Ennis, N.A. Gutierrez, and I.S. Maddox, *Proc. Biochem.*, 1986, **21**(5), 131
[5]   EEC Council Directive, 13.6.1988 (88/334/EEC).

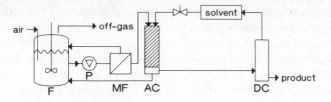

F = fermentor
P = pump
MF = microfiltration unit
AC = adsorption column
DC = distillation column

Figure 10 *Fermentation with on-line adsorption*

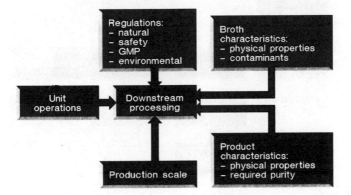

Figure 11 *Factors influencing downstream processing design*

	K	N	d $\frac{20}{20}$	Visc. cP	Bp^{760} °C
Hexane	5.5	5	0.66	0.33	68.7
Hexene–1	7.0	4	0.67	0.30	63.3
Toluene	10.3	4	0.87	0.58	110.6
Cyclohexane	6.0	5	0.78	0.94	80.8
MTBE	27.8	2	0.74	0.62	55.0
Methyliso–butylketone	21.3	2	0.81	0.59	115.9

N = number of extractions (ratio 1:4) to reach
    98% extraction efficiency

Figure 12 *Extraction efficiency for 2(3)-methylbutyric acid*

possibilities considerably and as a consequence has a large impact on both
the required equipment and the cost. From the Table given in Figure 12, it
can be seen that MTBE is a far more efficient solvent for the extraction of

methylbutyric acid than, for example, hexane. Unfortunately, MTBE is not on the positive list of allowed solvents.

Most flavour ingredients are volatile components excreted from the cells into the medium. This facilitates downstream processing considerably, avoiding the need for cell disruption. However, products contributing particularly to taste are quite often non-volatile and not excreted, like amino acids and taste enhancers. These products have to be isolated via a process of cell harvesting and disruption or lysis. An example of a biotechnological process whereby the interesting flavour compounds are localized in the cell is our process for the production of High Taste Enhancement yeast speciality (Figure 13).

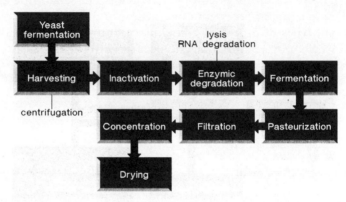

Figure 13 *High taste enhancement yeast speciality*

4   Unit Operations in Downstream Processing

In Figure 14, a survey is given of the most important unit operations used for the isolation of volatile flavour ingredients from fermentation broths.

Solids removal	Filtration Centrifugation Microfiltration	Purification	Distillation Crystallization
Biopolymer removal	Ultrafiltration	Drying	Conventional Spray Freeze
Isolation/ concentration	Extraction Distillation Hyperfiltration Chromatography		

Figure 14 *Unit operations in downstream processing of flavour ingredients*

Flavour Houses traditionally have built up considerable expertise and facilities for the isolation, concentration, and purification of volatile flavour ingredients. These same technologies can play an important role in downstreaming of microbiologically-produced ingredients. Since there is often considerable flexibility in the design of DSP, it is only sensible from the point of view of cost that the initial approaches should use existing technologies.

4.1 Centrifugation

Centrifugation[6] can be used not only for cell harvesting, but also for clarification of a fermentation broth prior to extraction, membrane filtration, or chromatography. Centrifuges can be very helpful in separating emulsions obtained in an extractive fermentation process or during solvent extraction. The different types of processes taking place during centrifugation can be classified by the physical phenomena occurring during separation (Figure 15). De-emulsification is the separation of an emulsion into its liquid phases; sedimentation takes place during a decantation process in which solids are separated from a liquid phase.

De–emulsification : L/L separation – extraction

Sedimentation   : S/L separation – decantation

Filtration   : S/L separation – decantation, washing

Figure 15 *Centrifugation processes*

Both phenomena are combined in the case of an extractive fermentation where, in one process step, the solids and the two liquid phases are separated.

Filtration takes place if the solids are sedimented on a filter while the liquids pass through. In this type of centrifuge the solid phase (filter cake) can easily be washed.

Examples

In our process for *High Taste Enhancement yeast specialities*, the yeast cells are harvested with centrifiges. The yeast wort is fed to the centrifuge. The yeast concentrate is washed and again concentrated by a second run through the centrifuge (Figure 16). Because of the high biomass level in the feed, we make use of nozzle bowl centrifuges in which the solids are continuously removed from the bowl.

---

[6] H. Hemfort, W. Kohlstette, Wesfalia Separator, Techn. Scient., Doc. no 5, 1988.

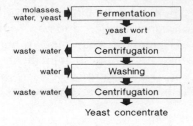

Figure 16 *Yeast concentrate production*

For the production of methylketones, there are two options after the extractive fermentation process (Figure 17):
1. Via steam distillation and fractionation, the pure ketones are produced.
2. By a centrifugation process, comparable to the well-known milk separation, the solids, the oil phase, and the water phase are separated in one run.

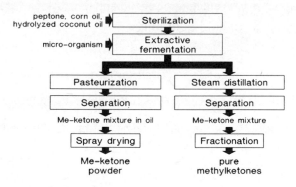

Figure 17 *Methylketones C_5, C_7, C_9, C_{11}*

In our process for γ-decalactone production, the fermentation broth is extracted after the cell mass has been removed by a micro-filtration or centrifugation step (Figure 18). During solvent extraction an emulsion is formed which does not separate easily. Again with the help of a centrifuge, a phase separation can be carried out effectively.

4.2 Extraction

Many flavour ingredients can be extracted from a solution in water by an appropriate organic solvent.[7] Recovery of the product and the solvent after extraction is done normally by distillation. Criteria for solvent selection are shown in Figure 19. The efficiency of the extraction process is, to a large

[7] P.J. Bailes, C. Hanson, and M.A. Hughes, *Chem. Eng.*, 1976, **83**, (2), 86.

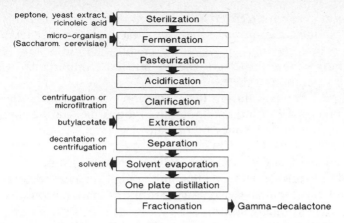

Figure 18 γ-*Decalactone*

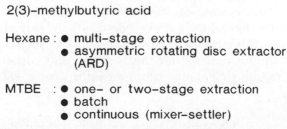

- high partition coefficient $K = \dfrac{C\ \text{org.}}{C\ H_2O}$
- immiscible with water
- density difference
- low viscosity

Figure 19 *Criteria for solvent selection*

extent, dependent on the partition coefficient of the product between the two phases.

In earlier discussions on the influence of regulations on downstream processing, the extraction efficiency of solvents for methylbutyric acid was exemplified (Figure 12). With hexane, a multi-stage extraction is required to achieve an acceptable extraction yield. In this case, use is made of an Asymmetric Rotating Disc extractor (Figure 20). If MTBE could be used, a one or two stage extraction would be sufficient to reach the same yield, and a simple batch extraction procedure could be followed.

2(3)-methylbutyric acid

Hexane : ● multi-stage extraction
　　　　 ● asymmetric rotating disc extractor
　　　　　 (ARD)

MTBE : ● one- or two-stage extraction
　　　　 ● batch
　　　　 ● continuous (mixer-settler)

Figure 20 *Influence of nature of solvent on choice of extraction equipment*

For large operations where preferably a continuous extraction procedure would be used, so-called mixer-settler extractors are available. The extraction solvents should be immiscible with water. This is particularly important with respect to solvent recovery. The *density* should be sufficiently different from that of the aqueous phase in order to facilitate a fast separation of the two phases after mixing. The *viscosity* preferably should be low. A high viscosity requires more dispersion energy, but more importantly would slow down the separation of the two phases.

Emulsion formation during the extraction process can be the cause of severe problems. Ultrafiltration to remove biopolymers or the application of de-emulsifiers can sometimes be of help. De-emulsification by centrifugation may also solve the problem.

Extraction with super-critical gas (CO_2) can have distinct advantages, as indicated in Figure 21. However, due to the high pressure required, a major investment in equipment is necessary. This is the main reason why this technology is not yet widely used.

Advantages : ● no residues, chemically pure
　　　　　　 ● low temperature
　　　　　　 ● non toxic
　　　　　　 ● low viscosity
　　　　　　 ● non–flammable
　　　　　　 ● odourless

Disadvantage: ● high pressures (>75 bar)
　　　　　　　 ● high investment cost

Figure 21 *Super-critical gas extraction*

4.3 Chromatography

For the purification of flavour ingredients liquid chromatography can be considered. In comparison with fractional distillation, it is often a more expensive approach but for smaller quantities, it can be the most effective technology to produce the right organoleptic quality. An interesting option for the isolation of flavour ingredients from a very dilute aqueous solution is reversed phase chromatography, as has been described by Di Cesare *et al*.[8] After adsorption the product can be eluted with ethanol. In order to circumvent blocking of the column, it is necessary to clarify the feed *e.g.* by ultrafiltration or steam distillation. In our production process for a malt flavour, we use chromatography for the isolation and concentration of the product. The resulting ethanolic solution can be used as such (Figure 22).

4.4 Distillation

Distillation, together with extraction, is the most commonly used technology in flavour ingredient isolation and purification. It is also of major importance for the downstreaming of volatile fermentation products. An advan-

[8] L.F. Di Cesare, R. Nani, and G. Bertolo, *Conservazione degli Alimenti*, 1987, **9**, 35.

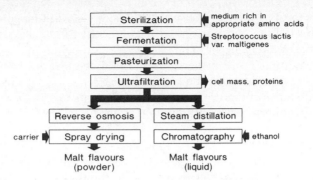

Figure 22 *Malt flavour*

tage for flavour ingredient producers is that, because the technology is so well developed and widely used, the required equipment is often available. A drawback of the technology is that relatively high temperatures are required. This can cause decomposition of thermolabile products. Steam distillation of the fermentation broth, either by direct steam injection or by indirect heating, can be an attractive first step in product recovery. After an extraction step, the solvent is normally removed by a simple batch distillation. For larger quantities a continuous approach can be attractive.

Final product purification is often done by fractional distillation. Since many flavour ingredients and their by-products are not completely thermally stable, fractional distillations are normally carried out under reduced pressure. In order to further reduce the risk of thermal decomposition, it is also advantageous to shorten the residence time at high temperatures. We make use of falling film evaporators in combination with columns packed with cascade mini rings, giving a high boil-up rate or with Sulzer packing, giving a very good separation efficiency.

4.5 Membrane Filtration

Membrane filtration[9-12] is an essentially non-thermal process used to remove cells or other suspended material, for the removal of biopolymers, and for the concentration of flavour ingredients (Figure 23). Removal of cell mass is often necessary before extraction with a solvent to circumvent autolysis of the cells during the extraction process which could give rise to formation of troublesome emulsions and off-flavours. Microfiltration (MF) is able to remove suspended larger particles like cells. Ultrafiltration (UF) is very suitable for the removal of larger molecules like biopolymers, whereas hyperfiltration or reverse osmosis is used for the concentration of smaller.

9 S. Ripperger, *Chem.-Ing.-Techn.*, 1988, **60**, (3), 155.
10 S. Ripperger and G. Schulz, *Bioprocess Eng.*, 1986, **1**, 43.
11 M. Cheryan, Nat. Res. Counc. Can. (Rep.) NRCC 29895, Adv. Reverse Osmosis Ultrafiltration, 1989, 443.
12 D. Mackay and T. Salusbury, *Chem. Eng., (London)*, 1988, **447**, 45.

Membrane type		Process type	Separation principle	Pressure
Microfiltration	MF	Cell removal/ clarification	Sieving	2 bar
Ultrafiltration	UF	Clarification/ fractionation	Sieving	3–10 bar
Hyperfiltration	HF/RO	Concentration	Diffusion	20–70 bar

Figure 23 *Membrane filtration*

molecules. In essence MF and UF are sieving processes in which the pore size of the membrane is the discriminating factor. Since in the case of large molecules the osmotic pressure is relatively low, only moderate pressures are required to make this process work.

In the case of hyperfiltration the pore size is so small that flavour molecules are retained, while the water molecules pass through the membrane. In this case, a high osmotic pressure is built up requiring a high pressure to operate the concentration process. This process can be better described in terms of diffusion rather than sieving.

Membrane processes are executed normally with cross-flow of the liquid over the membrane. The shear and lifting forces produced by the cross-flow largely prevent the formation of a layer of separated particles. Cross-flow filtration greatly facilitates maintenance of a reasonable flux through the membrane (Figure 24). A large variety of membrane types is available based on organic polymers, having different pore sizes, retention, and flux characteristics.

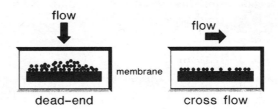

Figure 24 *Cross flow filtration* vs. *dead-end filtration*

Membrane modules basically can have the following configurations:
    flat sheet
    spiral-wound
    tubular and hollow-fibres.

A *flat sheet configuration* is rather versatile because it is quite easy to exchange sheets. A disadvantage is that the module is physically large. To realize sufficient shear and turbulence, screens are often used between the sheets, however these can lead to considerable pressure drop. This system is limited to rather low viscosity and low cell mass feeds.

Tubular and hollow-fibre membranes (Figure 25) are normally built into easily exchangeable cartridges. The smaller the diameter the less solids can be handled but the greater is the flux. Hollow-fibre modules are generally not autoclavable or steam-sterilizable and should be chemically sterilized. Tubes are particularly applicable for highly viscous and high cell density liquids.

Spiral-wound filters (Figure 25) also utilize cartridges and offer large areas in small packages. Because screens are used between the membranes, there are limitations in viscosity and solids concentration that can be used.

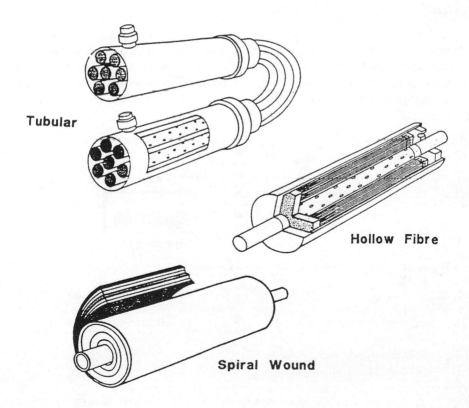

Tubular

Hollow Fibre

Spiral Wound

Figure 25 *Membrane module configurations*

Selection of the right membrane and configuration requires careful consideration of all possibilities and requirements.

In our process for the production of methylbutyric acids a combination of these techniques is used for the downstreaming of the fermentation broth. Methylbutyric acids are produced by microbiological oxidation of the corresponding alcohols (Figure 26). After acidification of the broth, cell mass and

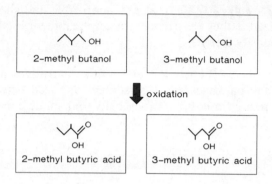

Figure 26 *2- and 3-methylbutyric acid*

biopolymers are removed by ultrafiltration. The clarified broth is extracted subsequently with hexane. Solvent recovery is effected by a simple batch distillation, after which the product is purified by a sequence of one-plate distillation and fractional distillation (Figure 27).

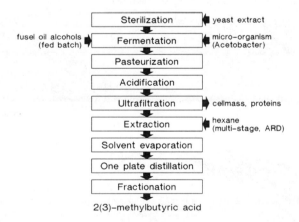

Figure 27 *2- and 3-methylbutyric acid*

Subject Index

DATE DUE

AUG 1 4 2003			